A NOTE ON THE AUTHOR

Louisa Preston is an astrobiologist and planetary geologist at Birkbeck, University of London; a UK Space Agency Aurora Research Fellow and unashamed lover of water bears. Her research focuses on places on Earth in which life is able to survive despite extreme conditions; such habitats provide clues on what alien life-forms might look like, and where we should search for them.

Having worked on projects for NASA and the Canadian, European and UK Space Agencies, the only thing Louisa enjoys more than devising ways to find extra-terrestrial life is writing and talking about it. She has published numerous articles and academic papers, and regularly appears on radio and television shows, such as the BBC's *The Sky at Night*. She is a TED fellow, and spoke about the search for life on Mars at the 2013 TED Conference.

Louisa has spent most of her life in one of two Londons – one in Ontario, and one in the UK – but she currently lives with husband, son and cat in Kent. *Goldilocks and the Water Bears* is her first book.

Also available in the Bloomsbury Sigma series:

GOLDILOCKS AND THE WATER BEARS

THE SEARCH FOR LIFE IN THE UNIVERSE

Louisa Preston

BLOOMSBURY
sigma

Bloomsbury Sigma
An imprint of Bloomsbury Publishing Plc

50 Bedford Square
London
WC1B 3DP
UK

1385 Broadway
New York
NY 10018
USA

www.bloomsbury.com

First published 2016. Paperback edition 2018.

British Library Cataloguing-in-Publication Data
A catalogue record for this book is available from the British Library.

Every effort has been made to trace or contact all copyright holders.
The publishers would be pleased to rectify any errors or omissions
brought to their attention at the earliest opportunity.

Library of Congress Cataloguing-in-Publication data has been applied for.

ISBN (paperback) 978-1-4729-2011-9
ISBN (ebook) 978-1-4729-2008-9

2 4 6 8 10 9 7 5 3 1

Illustrations by Samantha Goodlet

Typeset by Deanta Global Publishing Services, Chennai, India
Printed and bound in Great Britain by CPI Group (UK) Ltd,
Croydon CR0 4YY

To find out more about our authors and books visit www.bloomsbury.com.
Here you will find extracts, author interviews, details of forthcoming
events and the option to sign up for our newsletters.

For Daniel and Renley ...

Contents

Preface

The tale of *Goldilocks and the Three Bears* has charmed generations of children the world over, and has been borrowed and scientifically woven into the title of this book again – hoping to charm, but also challenge, its valued readers, encouraging you to think a little differently about the world and indeed the Universe in which we live.

Parents and teachers alike have used the fictional narrative of a plucky young girl and her cheeky invasion of the home of a family of bears to develop the imagination and story creation skills of children, as well as to convey levels of acceptable behaviour and manners, respect for personal privacy and respect of other peoples' property or belongings. And yet it contains another message that hopefully this book will inspire its readers to consider when imagining life out there in and beyond the *final frontier*: the idea that something has to be *just right* for it to be useful. The chairs tried by Goldilocks were either too big, too small or just right, the porridge was too hot, too cold or just right and the beds were too hard, too soft or just right. Only when each condition was *just right* and therefore acceptable to Goldilocks was the porridge eaten or the chair sat in or the bed used for sleeping. Just as in the search for life in the Universe, Goldilocks was searching for conditions that were perfect for her and so are we. Hunting across the *right* part of the Galaxy for planets and moons that are the *right* size, orbiting around the *right* type of star at just the *right* distance to be able to keep water as a liquid on their surface. Astrobiologists and Goldilocks have much in common.

In this book we shall take a tour of the biological Universe, exploring what life is made of, what it needs to originate and thrive, how resilient and adaptable it can be and how conditions do not actually have to conform to our ideas of what is or is not *just right* for organisms to survive and prosper. In many areas throughout the Solar System and beyond, conditions cover a multitude of extremes – too hot or too cold, too acidic or too alkaline, too dry or too wet, or too light or too dark – and in each of these on Earth, life has found its own version of *just right* so it can survive. The Goldilocks story of life is warped and stretched as life finds a way, no matter the challenges. If this is the tale on our planet, then why should it not be the same on other planets and moons, in other galaxies, and throughout the Universe? We may not be alone in the darkness of space for much longer.

'What about the Water Bears?' I hear you ask. Well, you'll have to continue reading to see where these little superheroes fit into this cosmic fairy tale …

CHAPTER ONE

A Brave New World

One of the oldest relationships on Earth is that between fact and fiction. It has not always been a happy one but over the centuries it has become apparent that one needs and inspires the other in more ways than we ever believed. The fact-loving subject of science is one of the most exciting disciplines in the world, although its teachings can get bogged down in boring educational texts and hidden behind fear, dogma and confusion. The analytical nature of science gives us the ability to perceive the anatomy of the Universe and every molecule in it, but it is the human imagination that gives it life. Descriptions of the history of the Earth, the stars sparkling across the cosmos, worlds so far away we may never see them and the wonderment of what life forms potentially exist out there, contain more exotic characters and magical realms than any fictional tale. By using storytelling, the walls of science are being broken down and a self-conscious reading public, aware of and excited by its own progress and the rapidly changing world around it, can finally engage with a subject that makes the existence of such entities entirely possible.

An Astrobiologist is Born ...

More than 100 years ago in a leafy countryside town just outside London, a 'gentleman' writes about the attempted destruction of the human race by Martians. A meteor lands on Horsell Common near Woking, Surrey, yet instead of a scorched lump of rock, the object opens – disgorging alien beings from the planet Mars. These Martians have abandoned their dry and dusty home during its last dying gasps and travelled to the living, breathing Earth in the hope of their salvation. With two large, dark-coloured eyes and a lip-less mouth, a big greyish bulk the size of a bear, glistening like wet leather, and brandishing two giant tentacles instead of arms, these monstrous creatures emerge on to the Earth – only to struggle to breathe in the oxygen-rich atmosphere and be forced to retreat quickly into their ships. They build frightening three-legged metallic machines with which to lay waste to the towns and villages west of London while forging their way to the capital. They extinguish everything and everyone in their wake, firing deadly heat-rays, spreading poisonous black smoke and a pervasive red weed. This tale ends well for humanity (spoiler alert!) as after a bitter struggle, and at a point where a Martian victory seems inevitable, a humble terrestrial microbe delivers the final blow. Bacteria, against which it turns out the Martians have no defences, infect and kill the invaders, ultimately saving humanity and the entire Earth.

England's father of science fiction, H.G. Wells, penned this story in 1897. Instead of instilling fear, this my first exposure to classic science fiction at a very young age made me curious. Was there actually life on Mars that could one day invade us? How could a tiny microbe kill the aliens when the 'great and powerful' humans could not? Add these questions to the fact that the entire story took place in my home county and completely annihilated the town in which I grew up, who wouldn't be hooked? This serial called *The War of the Worlds* saw H.G. Wells write the first alien invasion story and create a legacy of distrust, yet also curiosity, surrounding the possibility of contact with alien life. He was one of a few in the nineteenth century who began to popularise the alien, both negatively and positively, and brought to the public arena the science behind the search for extraterrestrial life and the environments that exist on other worlds. This story and others like it have straddled the boundary between science fact and science fiction, inspiring generations of scientists to pursue space science, including me, and paved the way for a new scientific discipline to be born: Astrobiology.

The Science of Alien Life

How does life begin and evolve from the simplest carbon-rich speck to a fully conscious individual? Is there life beyond Earth and if so, how can we find it? What is the future for life on Earth and its existence in the Universe? Humans have looked up at the night sky for thousands of years and asked these questions. We have gazed at the myriad stars twinkling in the darkness, wondering if somewhere out in the silence of space there was anyone staring back. Everyone has probably at some point pondered where life on our planet came from or if we are alone. Today, astrobiology works tirelessly to address the compelling mysteries surrounding extraterrestrial life, while embracing the study of the origin, evolution, distribution and future of life on this planet. It is an ever-evolving eclectic multidisciplinary field, encompassing a range of subjects including physics, chemistry, astronomy, biology, molecular biology, ecology, planetary science, geography and geology – all working together to investigate the possibility of life on other worlds. Yet Astrobiology is an enigma of a science. Unlike geology, which arose to describe the physical materials and origins of the Earth, or microbiology to explain observations of life under the microscope, astrobiology has yet to prove its subject matter actually exists – alien life is currently still fictional. Instead it is all about the hunt, the discoveries, the dead ends and the challenge to think outside of our habitual comfort zone and dare to imagine other forms of life and where they might be hiding.

It was only 50 years ago that humanity began to extend its presence into space – first with robots, then with animals and finally with humans. This tentative expansion of our species towards other worlds has been made possible by the development of technology, which has finally started to reach a level that can complement and support our imagination and desire for exploration. However, considering the size of the cosmos and the growing number of promising sites on many worlds where life might quite like to snuggle up, the search has barely begun. When we finally find life on another world – and we will – it will be one of the most significant cultural events in human history, having a profound resonance on the question of our origins. It is not surprising, therefore, to find that such possibilities have been discussed by every human civilisation and culture, primitive or advanced, as far back as we have written records. Even before these thoughts were given a name and embraced within a scientific discipline of their own, such extraterrestrial wonderings found their outlet through myths and fairy tales, conversations, cave paintings, lectures, letters, fictional literature, philosophy and religions, music and poetry, then later through films and TV shows and video games. Today, astrobiology also features strongly on social media via personal and group blogs, Twitter, Facebook and YouTube.

Star-crossed Lovers

The marriage between science fiction and science fact is built on truth. For instance, science fiction frequently contains the theme of humans living on an alternative planet (for better or for worse) or escaping from Earth altogether and settling on other planets, moons or asteroids in the Solar System, and even star systems beyond. Science fact forecasts that by the end of the twenty-first century the Earth's human population is predicted to surpass 11 billion individuals, crowding the biosphere and further depleting the raw materials on which life depends. In this all-too-likely version of the future many questions will surround the sustainability of the global population, the growing pressures on the natural environment, world food supplies, and energy resources. Ally this to the fact that what is available now is being polluted and quite possibly ruined for the next generations, and the future does not look promising. Science fiction may one day need to become science fact.

For the vast majority of people, the idea of extraterrestrial life entered their lives not as science but as fiction. Even scientists, whose reputation implies an innate love of the technicalities of scientific concepts, are commonly influenced by alien literature and film. The concept of alien life forms is deeply felt and firmly ingrained in the human mind, and has culminated in some of the most popular movies of all time, including *2001: A Space Odyssey* (1968), *Close Encounters of the Third Kind* (1977), *Star Wars* (1977 to present), *Alien* (1979, 1986, 1992 and 1997), *E.T. the Extraterrestrial* (1982), *Independence Day* (1996 and 2016), *Contact* (1997), *District 9* (2009), *Avatar* (2009), *Guardians of the Galaxy* (2014) and *The Martian* (2015).* In fact, if you search for films that include a theme of extraterrestrial life you find that on average over 10 movies a year are released. This demonstrates that a beautiful symbiotic relationship exists between art and science, bringing together creative thought and scientific theory, driving forwards both innovation and exploration. As such a number of astrobiology-rich stories over the years have amassed quite a cult following due to their gripping

* Yes Mark Whatney is a human but whilst living on Mars he is the alien.

storylines, relatable characters and faithfulness to the science, inspiring generations of the public and scientists alike.

42

What is the answer to the Ultimate Question of Life, the Universe, and Everything? According to legendary comedy/sci-fi author Douglas Adams (with whom I incidentally share a birthday), the answer is 42. In 1979, he published a novel based on his radio show *The Hitchhiker's Guide to the Galaxy*, the first in a trilogy of five books. Yes, a trilogy of five. *The Hitchhiker's Guide* follows the story of a hapless human called Arthur Dent, who is saved from Earth's destruction with just seconds to spare by his good friend Ford Prefect. Prefect, whom Dent at first believes to be human, actually turns out to be an alien who named himself after the Ford Prefect car in a bid to blend in with what he assumed to be the dominant terrestrial life form. He is working for the Hitchhiker's Guide to the Galaxy – a combination travel guide/Wikipedia for intergalactic travellers roaming the Universe by grabbing rides on passing spacecraft. The planet Earth is actually a computer, just mistaken for a planet because of its size and use of biological components, and is destroyed by the Vogons to make way for a hyperspatial express route. Dent and Prefect end up on a series of perilous adventures aboard a ship stolen by the President of the Galaxy, Zaphod Beeblebrox, along with another human runaway Trillian and a depressed robot, Marvin the Paranoid Android.

The importance of science in this famous story is stated right at the start: 'Far out in the uncharted backwaters of the unfashionable end of the western spiral arm of the Galaxy lies a small unregarded yellow sun. Orbiting this at a distance of roughly ninety-two million miles is an utterly insignificant little blue green planet.' This hypothesises how unimportant we are and how wrong humanity is about the Universe. In that way, this very much fictional story is like science in real life: it forces us to reconsider and challenge what we think we know and makes us feel small and insignificant at the same time. Exploration in *The Hitchhiker's Guide to the Galaxy* is about moving out into the unknown and experiencing new things. This story opened people's minds to a new way to view the Universe and ignited a wish in many to explore the Galaxy. It gave an insight into the possible types of aliens out there and the role of artificial intelligence – quite possibly the next step in our evolution. All the while, a ribbon of comedy keeps it light and engaging. The idea that the answer to any question is 42, and the positive impact the story has had on so many, is visible throughout today's culture.

Seeking Out New Life and New Civilisations

Ever since the starship *Enterprise* first warped across television screens in 1966, *Star Trek* has continued to inspire audiences with its portrayal of a future space-faring species game to 'boldly go where no [one] has gone before'. Its creator Gene Roddenberry, and his daring writers, started with real-life science and s-t-r-e-t-c-h-e-d it to create amazing imaginative inventions and intricately crafted worlds that could fit within remotely plausible storylines. They stayed true to the map of the Universe and kept the stars realistically far away, but designed the human species with the power to reach them in days or weeks, instead of in lifetimes. To reach the nearest star system today, Alpha Centauri, it would take the Space Shuttle and current technology up to 165,000 years. Needless to say, those who start the journey, and even their great-grandchildren, would never see its end. Generational ships such as the *Enterprise* are something commonly thought about for future space-faring humans. *Star Trek* (and its subsequent spin-off series and films) also gave us magical devices such as the transporter, advanced medical instruments and the holodeck, which were included to showcase the fantastical tools that might be built by human engineers in a future where humanity has technologically progressed into the realms of current science fiction. *Star Trek* gave the public a vision of what may one day be possible, and that's just one reason why the shows have been so popular. The real science lies in an effort to be true to humanity's greatest achievements while making it accessible and entertaining to watch. Obviously it cannot be wholly faithful to science and technology today as it is virtually impossible to create a perfectly accurate science-fiction TV series. Some scientists of course discredit *Star Trek* because of certain scientific errors or impossible events, but this is unfair. Accurate science is seldom exciting and spectacular enough to base a weekly adventure TV show upon. More faithful to science than any other science-fiction series ever shown, *Star Trek* has attracted and excited generations of viewers about advanced science and engineering and it is one of a select few shows that depict scientists and engineers of both genders and all races positively, as role models and people to aspire to.

A huge branch of astrobiology focuses on the future of human civilisation and the design of tools and technologies to allow us to one day live among the stars. Science inspired countless aspects of *Star Trek* and now many of the tools designed for the show are inspiring science. Remember the Replicator – a magical aperture in the bulkhead where you could order anything you desired? Although we cannot materialise food out of thin air (wouldn't that solve a lot of problems), we can create many other useful objects. In December 2014, NASA emailed the design plans for a wrench up to an astronaut on the International Space Station (ISS), who was then able to create it physically using a 3D printer. Unlike science-fiction Replicators, which seemingly produce objects on demand and whole ('Tea, Earl Grey, Hot'), this 3D-printed version came out as a 20-piece assembly kit that Astronaut Barry Wilmore then snapped together into a working socket wrench. We can also 3D-print artificial organs tailored to fit individual people, which are meant to replace or even enhance human parts, such as titanium replacement hip joints and made-to-order polymer bones that can be used to reconstruct damaged skulls and fingers. The Replicator is not as far-fetched as you may think.

What about artificial intelligence, the potential next step in evolution? Although there has been a lot of press about a computer program that reportedly passed the Turing Test by fooling judges into thinking it was human (it did not), no one seriously suggests that this particular program is sentient. It is still definitely an *it* instead of a *he* or *she*. Sentient machines remain in the realm of fiction, yet the dream of creating a Lieutenant Commander Data is very real. The ways in which thoughts are encoded and transmitted within the human brain remain only crudely understood, preventing real intelligence from being developed just yet. However, simple brain-to-machine commands can be transmitted and have been created, enabling impaired or paralysed people to control prostheses and machines. Do you recall *Star Trek*'s Universal Translator? Microsoft recently announced Skype Translator, which allows near real-time audio translation from one language to another, utilising advances in speech recognition and machine translation technologies. Currently in its infancy, it isn't a galaxy far far away* from reality. Interestingly, the Apple iPad bears a remarkable resemblance to hand-held touchscreen networked computers called PADDs (Personal Access Display Device) used by some Starfleet Officers in *Star Trek*. And although no medical tricorders have been built yet, we do have handheld sonograms that help to observe a foetus and organs inside the body.

The human imagination can create a wealth of gadgets that future generations may find useful but where it really excels is in creating life itself. Now that we understand biochemistry a little better, most scientists agree that life probably exists

---

* A *Star Wars* rather than *Star Trek* reference.

somewhere out there in other solar systems. The chemical elements needed for carbon-based life are common in the Universe, so maybe life forms built like us are numerous in the Galaxy, too. Our imagination can conjure up all kinds of creatures with any number of arms, legs, eyes or even heads that are quite often equally or a lot smarter than we are. Yet it seems doubtful that humanoid shapes would be as common as the alien races depicted in science-fiction series such as *Star Trek*. We have to allow the show some concessions here, as there are only so many shapes and sizes actors come in. Could half-human/half-alien hybrids such as Spock ever exist? It seems almost impossible, although scientists and nature alike have already created interspecies hybrids from plants, the Killer Bee and Wholphins (false killer whale × bottlenose dolphin). Spock is not totally beyond biochemical reality, but definitely at the fictional edge of science fact.

I See You

Next we come to one of the most scientifically plausible, almost educational, astrobiology-themed films of all time, and it is highly enjoyable to boot. *Avatar* (2009) is set in the year 2154 and humankind has reached out to the stars. The white-yellow glow of Alpha Centauri A – a star very similar to our Sun – illuminates the Saturn-sized gaseous giant planet Polyphemus and its fifth moon, the tropical world of Pandora. On this lush green satellite, great beasts roam the jungles and pterodactyl-like creatures soar through the sky. A sentient, blue-skinned humanoid species known as the Na'vi lives here, in harmony with nature and at war with humanity.

Its creator James Cameron, a physics major himself, created this world using science. He set Pandora and Polyphemus in the real Alpha Centauri system, the closest stellar family to Earth. This system actually has three stars, all revolving around one another. At its heart lies a pair of sun-like stars: the first, Centaurus A, is some 20 per cent larger than the Sun, while the second, Centaurus B, is 15 per cent smaller. A third star, Proxima Centauri, orbits them both and is a red dwarf that is 80 per cent smaller than the Sun. Since both Cen A and B are chemically similar to the Sun, the same general mixture of elements that allowed life to develop on Earth should have been available in the primordial soups of both their planetary broods. While planets in other solar systems make the news on an almost daily basis, one of the most remarkable announcements occurred in 2012 when astronomers claimed the discovery of an Earth-like planet circling Cen B, a mere 4.3 light years away. In astronomical terms, that is almost close enough to touch. This discovery may now prove to be false due to a 'ghost' in the measurements taken, but it hinted that perhaps Cameron had the right idea and a world like Pandora won't exist solely in the realms of science fiction and fantasy for much longer.

Making Pandora a moon is a wonderful acknowledgement of recent science. Astronomers in the first instance are looking for planets like Earth – small and rocky – within the 'Goldilocks zone' of their star: a not-too-hot, not-too-cold orbital band around a sun where life-giving water can be liquid on a planet's surface. Nonetheless, small planets, like the Earth, are hard to locate and scientists have found many more of the larger gas giants such as Saturn in the Goldilocks zone around other stars. Those planets are not habitable by life as we know it – although their moons, should they have them, could be. Icy and rocky moons surround all the gas giants in our Solar System, so it is feasible to suggest that this may be the case around alien Jupiters and Saturns. There may be Earth-sized rocky moons orbiting another gaseous world not too far away. A huge problem for life on one of these moons would be extreme radiation emitted from the gas giant. For instance, the daily radiation engulfing Jupiter's moon Io is 4,000 times the lethal dose for humans. Science allows James Cameron to give Pandora a life-protecting shield in the form of a robust magnetic field created by the moon's superconductive rocks, which deflects this harmful radiation and allows life to thrive on its surface. To highlight this, at one point in the film, a spectacular aurora is seen dancing overhead, as happens around the North and South Poles when solar winds hit Earth's magnetic shield.

Movies such as *Avatar* are challenging as they get us thinking about the possibilities, and especially the physical attributes of life. The moon Pandora is rocky, rich in the fictional mineral Unobtanium (an old in-joke in science fiction for materials with physically impossible qualities). From orbit, it looks a lot like Earth, with vast blue oceans, and continents covered in lush tropical rainforests, suggesting that both it and its mother planet must lie fairly close to their Sun, taking advantage of its light and warmth. Pandora's atmosphere is mainly composed of nitrogen, oxygen, carbon dioxide and xenon, and is 20 per cent denser than the Earth's atmosphere. The carbon dioxide makes it toxic to humans, acknowledged in the film by the use of oxygen masks. The trees on Pandora resemble those on Earth in colour and similarly have trunks, branches and leaves, although owing to Pandora's lower gravity (20 per cent that of Earth), the shapes and proportions have been able to be exaggerated. Also because of this lower gravity, most creatures are hexapods (with six legs) and are often huge. The Na'vi, however, are strikingly reminiscent of humans, which as mentioned before is a common science-fiction theme. Their bodies, however, are larger, allowing them to survive in the lower-gravity conditions, and are built to hunt, being blessed with vision beyond the visible range of that of humans, feline ears, a tail for balance, and a snout. All these adaptations are for greater sensory perception, which aids their survival. The Na'vi's skin is blue, containing a pigment of cyanin, producing colours in the blue, purple and cyan spectrum. Na'vi blood is red, utilising an iron compound similar to haemoglobin to transport oxygen throughout the body. Bioluminescent skin cells emit light when ambient light levels are low.

All this detail may seem extreme when used solely for the creation of a fictional world and species, but it's the detail that gives the story its strength and credibility. Cameron and his team of advisers built this entire world upon the terrestrial laws of physics, biology and chemistry. The characteristics of the moon and the physical adaptations of its inhabitants are based upon detailed observations of other planets and moons, and the adaptations to different conditions seen in Earth-life, and common evolutionary trends. This film has predicted in detail what life, and its environment, might look like on alien planets and moons.

Extraterrestrials in most science-fiction stories may not be depicted as perfect human replicas, but do appear as pseudo-hominids that are eerily familiar. They tend to share our bipedal locomotion; bilateral symmetry; build of head, trunk (body), two arms and two legs; upright posture and forward-facing, stereoscopic eyes. Even robots are designed in our image to an extent. The reason lies in science: designs are commonly inspired by the life that we see around us and, most importantly, can interact with. As such, science fiction also commonly assumes alien life will be complex and intelligent, because an immobile silent blob of wobbling plasma would not exactly make for gripping viewing. However, extraterrestrial simply means 'beyond Earth' and there is nothing that says this beyond-Earth life has to be humanoid in form, especially considering that more than 90 per cent of life on Earth looks nothing like us.

Rock, Paper, Scissors, Lizard, Spock

Television shows such as *The Big Bang Theory* bring together the science, the story and the people. Fictional stories based on fact have communicated science to a wider audience for centuries with both real and fictional scientists themselves proving to have an even stronger impact. *The Big Bang Theory* is a California-based comedy that follows a group of self-professed geeks including a NASA engineer, an astrophysicist and two particle physicists that has made science chic again and is even credited with consolidating the growing appetite among teenagers for the once unfashionable subject of physics. Walking the narrow tightrope between science and sitcom, this show is beloved by critics, audiences and scientists alike for its quick wit, incredibly geeky yet relatable characters, and its science and science-fiction storylines. In more than one episode, there is a nod to astrobiology, most famously when the NASA engineer Howard crashed the Mars Exploration Rover into a crevice while trying to impress a girl. The data that it sent back contained the first clear indications that there may have been life on Mars, but he could not take the credit. The show not only embraces science and cult followings of many science-fiction shows, but also the geeky stereotype of scientists themselves while introducing a multitude of scientific concepts. Sheldon at one point refers to geologists like myself as the 'dirt people' – and it is not a compliment – but I'll let him off this once.

The First Astrobiologists

These twenty-first-century writers and scientists are not the first to wonder about the stars and envisage what alien life might be like (although they shout about it the loudest), and they will not be the last. As far back as 4,000 years ago the ancient Babylonians observed and recorded the position and movements of stars, planets, the Moon, comets and eclipses as seen crossing their night-time desert skies. At around the same time on the other side of the world, the ancient Chinese were also beginning to document the heavenly bodies above them and early Hindu writings in India showed a similar trend. Interest in the existence of life really awakened with the ancient Greeks. Democritus, who lived from 460 to 370 BC, is considered by many to be the father of modern science and potentially has the great honour of being the first astrobiologist. He proposed that originally the universe was composed of nothing but tiny atoms that collided together to form larger units – including the earth and everything on it. He even theorised about exoplanets – distant planets and moons spinning around other suns. Democritus said that some planets would be arid and lifeless, while others would bear life similar to, but not necessarily identical to, that on Earth. His teachings, written over 2,000 years ago, actually align with what we think and are starting to observe today. A contemporary of Democritus was Aristotle (384–322 BC), who is much more commonly remembered today (as was his mentor Plato, c. 428–348 BC), but who firmly rejected the idea of life existing anywhere other than on the Earth.

Thankfully, by the last century BC, Ancient Roman thinkers such as the poet and philosopher Lucretius (c. 99–55 BC), disagreed with Aristotle, and expressed in writings their belief in other inhabited worlds. In the fifteenth century CE, the work of Copernicus, and in the sixteenth century Galileo with his telescope, finally put the Sun where it belongs: at the centre of the Solar System. This coincided with the Renaissance, a rebirth of Western science and culture, and was a precursor to the Enlightenment of the seventeenth and eighteenth centuries, in which rationality was preferred to tradition and untested beliefs. In the early eighteenth century, many gentleman scientists, including Sir Isaac Newton – most famous for his love of apples and ideas about a thing called gravity – speculated on the existence of other worlds.

Despite this new phase of illumination, the belief in 'cosmic pluralism' or the plurality of worlds still existed – a non-science-based faith conceived in Ancient Greece, which held that numerous worlds in addition to Earth may exist and harbour extraterrestrial life. In 1862, Camille Flammarion wrote La pluralité des mondes habités (Plurality of Inhabited Worlds), a factual account of the philosophy of the possibility of life on other worlds. The creation of the telescope, instead of helping to dispel this idea, actually appeared to prove to many that a multitude of worlds containing life was a reasonable assumption. As greater scientific scepticism and rigour were applied to the question, it ceased to be simply a matter of philosophy and theology and became influenced and educated by astronomy and biology. Fiction started to be based upon and drawn from fact.

Up to this point, a number of prominent 'gentlemen' have dominated the story of both factual and fictional astrobiology, yet there are some truly epic women who have also driven space science and astrobiology forwards, and I hope you don't mind me highlighting it here. They may not be as well known as Marie Curie or Ada Lovelace, but without them science would not be where it is today. There is a reason when looking back at history, of course, for why we are not as familiar with their efforts – Western women have only been allowed to study science at university since the late 1800s and historically have faced a great struggle to be able to engage with scientific subjects. Even though it was not easy and records are few, there is a number of brilliant and influential females who have had a hugely positive impact on mainstream science, fact and fiction – from classic lecturers and professors to scientific researchers, teachers, authors, TV presenters, science journalists, writers and communicators.

Caroline Herschel (1750–1848), sibling to the famous astronomer Sir William Herschel, assisted her brother in his observations and in the building of telescopes and became a brilliant astronomer in her own right, discovering new nebulae and star clusters. She was the first woman to discover a comet and was the first British woman to be paid for her scientific work. She continued her astronomical studies until her death at age 97, compiling a catalogue of nebulae and increasing the number of known star clusters from 100 to 2,500. Mary Anning (1799–1847) was the first female fossil-hunter, and recovered the first ichthyosaur from a seaside cliff near Lyme Regis, England, when she was around

11 years old. In addition, she found long-necked plesiosaurs, a pterodactyl and hundreds, possibly thousands, of other fossils that helped scientists draw a picture of the marine world 200–140 million years ago. There was also Mary Somerville (1780–1872) who experimented with magnetism and produced a series of writings on astronomy, chemistry, physics and mathematics. Other great women, such as Rosina Zornlin, Daphne Jackson, Jocelyn Bell Burnell, Williamina Flemming, Cecilia Payne-Gaposchkin, Joan Feynman, Lydia Becker, Margaret Gatty, Mary Ward, Agnes Giberne, Agnes Clarke and Eliza Brightwen – to name but a few – pushed science and society forwards, and in turn paved the way for astrobiology to evolve. Today a huge cohort of intelligent, driven, and creative women proudly carry on their legacy.

This brings us up to the present, and the earliest published use of the word 'astrobiology'. This is credited to (although not everyone agrees) an article written by Lawrence Lafleur of Brooklyn College in 1941 that quite rightly described it as 'the consideration of life in the Universe elsewhere than Earth'. The science of astrobiology in the twentieth and twenty-first centuries has laid significant groundwork for the understanding of the genesis and evolution of life in the Universe. Exploring the farthest reaches of the Earth has uncovered fossils, organisms, and ecosystems that have all led to significant insight into the early Earth – possible models for life's origins, as well as a huge expansion of the recognised environmental limits of life. Laboratory work, coupled with astronomical observations, has added another significant piece of the puzzle and space missions are finally taking astrobiologists away from Earth and out into the cosmos. Given the timeless fascination with questions of the origins and prevalence of life, the science of astrobiology will surely endure long into the future.

Birth of the Alien

One of the most popular themes at the heart of astrobiology and nearly all space-based science-fiction epics is that of the extraterrestrial, the alien, or the menace from space. It is a topic that has greatly captured the public's imagination. Despite the long history of speculation surrounding life in the cosmos, the image of an alien only entered the realm of public literature in the last third of the nineteenth century, even though the topic had already inflamed the popular imagination back in the seventeenth century. The birth of the modern-day alien seems, therefore, to be intricately linked to advances in astronomy and the rise of the theory of evolution.

The alien, it turns out, was invented independently three separate times: in France, Germany, and England, spurred on by the imaginative science of an American. The people most often hailed as its creators are Jules Verne in France, Kurd Lasswitz in Germany and H.G. Wells in England. The famous novelist Jules Verne first discussed extraterrestrials in his 1870 novel *Autour de la Lune* (Around the Moon), but went no further as he constrained his imagination with science. He never wrote a story focused solely on aliens, as he needed proof of their existence first. In 1897, Kurd Lasswitz, the father of German science fiction, published *Auf Zwei Planeten* (On Two Planets), in which intelligent advanced Martians travelled to Earth, not out of some dire need to escape a dying planet or to colonise and rule the puny Earthlings but simply out of curiosity and a thirst for exploration. He believed that using aliens in his story could help illuminate the important role that science and technology played in society. He was not a scientist himself but a philosopher and historian, who adopted a scientific evolutionary universe in his stories. He once wrote that the natural order of the Universe demands not only that 'living, feeling, thinking creatures' exist on worlds currently inaccessible, but also that there should be 'infinite gradations of intelligent beings inhabiting such worlds'. Also in 1897, H.G. Wells serialised *The War of the Worlds* – a story full of scientific ideas gleaned from evolutionary biology and astronomy,

undoubtedly influenced by Wells's mentor, the biologist
T.H. Huxley (1825–1895). He devised the look and
intelligence of the Martians, explored how they would
have evolved beyond us due to Mars' greater age; their
ability to travel through space propelled by hydrogen gas,
the means by which they coped with Earth's stronger
gravity; and finally their demise due to disease. This book
launched not only Wells' career and legacy, but also that of
the alien itself. Its subject matter created an emotional
connection with its readers while exploring many, some
may say mundane, scientific facts.

The spread of the ideas of extraterrestrial life has jumped
from science to literature to film and back again. Although
the subject has always been popular since it first burst onto
the literary scene, of all the rich variety of science subjects
available it could never have been foreseen that this
completely fictional, currently unproven, subject matter
would become such a universal theme of popular culture,
the poster child for science fiction itself – and no life form
is more famous than the Martian.

Civilisations on Mars

Astrobiology is so much more than the search for life on
our nearest neighbour. Yet the story of this particular
search is one of the oldest and has inspired more science
writers than any other world. Mars has been a favoured
setting for the alien throughout the millennia, and the
leading candidate in the *real-life* search for extraterrestrial
beings. Known to exist since ancient Babylon, closer
observations of Mars only began in the 1500s, with Danish
astronomer Tycho Brahe (1546–1601). It was the Dutchman
Christiaan Huygens (1629–1695) who first speculated
about life on the planet. Extraterrestrial life, he said, is
not ruled out by the Bible, and why create other planets
if not to populate them? His ideas got a boost in the
eighteenth century by Sir William Herschel (1738–1822),

who, believing the dark areas on Mars to be seas, wondered if living Martians might not 'enjoy a situation similar to our own'.

Over the course of the twentieth century, the increasing power of telescopes and their detectors, the development of photography, photometry and spectroscopy, and the ability to send spacecraft to Mars, has further fuelled the search. Twentieth-century science has progressed from a hypothetical question of intelligence in the Solar System, to an exploration into the possible existence of vegetation and water, and finally to searching for organic molecules, fossil life and microorganisms themselves. A central character in this saga is Percival Lowell (1855–1916) and his startling idea that linear 'canal' features he saw on the surface of Mars were proof that the planet was inhabited. He was not the one to discover these – that crown is placed upon the head of the Italian astronomer Giovanni Schiaparelli (1835–1910) in 1877 – but Lowell made it a public sensation in 1894. Schiaparelli fuelled Lowell's theory by describing Mars as a planet of change, with two polar caps composed of snow and ice that through melting produced a temporary sea around the northern cap. He believed this water was distributed over great distances by 'a network of canals, perhaps constituting the principal mechanism (if not the only one) by which water (and with it organic life) may be diffused over the arid surface of the planet.' When Lowell joined the debate about the possibility of life on Mars, the general opinion was that the canals were cracks in the Martian crust made during the formation of the planet, although some argued for their artificiality. Backed by his comfortable financial situation, Lowell's passion for astronomy and interest in the notion that the planet might sustain life drove forwards publication of three major works between 1895 and 1908 and the construction of the Lowell Observatory – the first to be sited at high altitude and remote from urban light pollution. He wrote in 1895 that 'a system of irrigation seems an absolute necessity for Mars if the planet is to support any life upon its great continental areas.' He did not just mean any old life, but intelligent life. Life that had evolved enough to control the usage and distribution of water on a drying planet. He and his staff catalogued 183 canals, basing their artificial nature on their straightness, uniform width, and systematic radiation from specific points. He also proposed that vegetation was growing on Mars, and that he was only able to see the canals as they were lined with strips of fertilised land. By 1895, this was a story fully in the public domain and, although outrageous even to contemplate today, science could not disprove or confirm it for more than 20 years. The instruments of science just were not ready yet and personal bias still reigned, driven by scientists with strong imaginations. We now can prove that Lowell's canals actually correspond to no physical surface features on Mars, except for one – the large canyon of Valles Marineris. Lowell was simply observing an optical illusion. He even noted very broad, streaky canals on Venus, again false. Cloud-shrouded Venus was dubbed the most Earth-like planet and even said to harbour intelligent life, yet the planet quickly fell victim in the extraterrestrial life debate, when in 1940 conditions were considered too harsh to support life in any form. Although Venus was abandoned, Mars has remained as a possible site for life – despite our current understanding of its environment. This in no small part is due to science fiction.

The Future of Astrobiology

The drive to gain knowledge is a fundamental and almost evolutionary characteristic of our species. It is not hard to believe that our exploration of the cosmos and under-standing of the how, why and uniqueness of our existence may well influence the future of humanity. Our current astrobiological voyage across the stars in search for life promises to alter profoundly and to expand our notions of life, its origins and its future. If extraterrestrial life were found to exist, if a second origin of life were discovered on another world, or if life very different from Earth life were found outside the realms of life-as-we-know-it, it would change the very nature of what it means to be human, both scientifically and personally. In the search for life in the Universe we are not only looking for life we recognise, but that which might have evolved to thrive in conditions outside the bounds of Earth environments – the *weird* life only previously seen in our imaginations and in science fiction. Despite the literal meaning of the word, astrobiology is so much more than the search for biology in space. It covers the origins, evolution and future of life in all its guises on every world. As such there is one other form of life to consider, one that does not yet exist but that may be created in the future. The destiny of astrobiology may lie within the ecosystem of the World Wide Web.

Several researchers have put forwards the following premise, originally presented by SETI Senior Astronomer Seth Shostak: 'That once a society creates the technology that could put them in touch with the cosmos, they are only a few hundred years away from changing their paradigm from biology to artificial intelligence.' Many scientists predict a strong artificial intelligence will have been developed here on Earth by 2050 – about a hundred years after the invention of computers, or a hundred and fifty years after the invention of radio communication. The dominant intelligence in the cosmos may one day end up not being biological. Can we replicate our own intelligence, or something similar, and is it actually a good idea? This topic has made it to the mass media multiple times. First there was Stephen Hawking, then Elon Musk, and most recently Bill Gates. All of these visionary people have suggested that artificial intelligence (AI) is something to be watched carefully, 'lest it develops to a point of an existential threat'. What this threat might be has yet to be established. Hawking has suggested that it might be in the capacity of a strong AI to *evolve* much, much faster than biological systems – ultimately gobbling up resources without a care for the likes of us. It will do what it must to survive (Darwin's 'survival of the fittest'), even if this is at the expense of its creators. A system with the complexity of a human brain is almost certainly needed to sustain true intelligence, and today we have to wonder if the Internet itself is capable of achieving this. Estimates suggest that there are about 15 billion active Internet connections today and by 2020 there could be 50 billion. It is a world that's increasingly populated by algorithms whose movements and decisions are in part inspired by biological intelligence, or at least our impression of it. Code talks to code, software talks to hardware. Influencing this ecosystem is the powerful and sometimes irrational human mind, with our likes and dislikes guiding the flow of information, seeking to understand what we might search for next, as individuals or as a population. Could something akin to a strong AI emerge from all of this?

This theme has, of course, been translated into numerous science-fiction blockbusters such as *Metropolis* (1927), *The Day the Earth Stood Still* (1951 and 2008), *2001: A Space Odyssey* (1968), *Blade Runner* (1982), *A.I. Artificial Intelligence* (2001), *I, Robot* (2004), *The Matrix* and *The Terminator* franchises (starting in 1999 and 1984), the plucky quadrilateral robot side-kicks in *Interstellar* (2014) and the incredible *Ex Machina* (2015). Even Pixar's cartoon *WALL-E* (2008) has taught children about AI and the possible future of humanity.*

As a subject, astrobiology is tackling one of the most profound questions we can ask about our place in the Universe – is the 4-billion-year-old biological experiment that is life on Earth unique? As Arthur C. Clarke once said, 'Two possibilities exist: either we are alone in the Universe or we are not. Both are equally terrifying.' The prospect

---

* WALL-E fell in love with fellow robot EVE, the Extraterrestrial Vegetation Evaluator, who is technically an astrobiologist.

and nature of life, both on and beyond Earth, is one of the few topics in science that can grab the attention of any public audience. Science fact and science fiction are working together to drive humanity's exploration of the Universe forwards and to support our quest to find a biological cosmic connection.

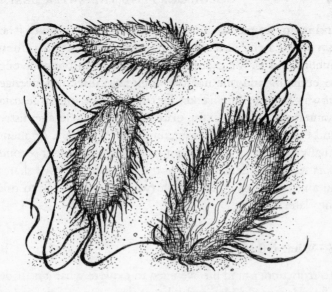

CHAPTER TWO
Life As We Know It

It may seem logical to start the story of the search for life at … well … the beginning. Yet it won't make much sense without first knowing what life actually is and what it needs to survive. It isn't pretty and is chemistry heavy, but provides a solid background to understanding what makes life possible. On our planet, living organisms have spread to every nook and cranny that can sustain them – in every direction you look; across land, water and air; from the driest deserts, balmy tropics, and tallest mountains to the coldest Arctic ice; on, within and under rocks and deep within nuclear reactors; there are even living beings inside the bodies of other living beings. Yet even with all this life, we still cannot quite explain it. The question 'What is life?'

seems so simple to answer. If something is alive, it is obvious – a chair isn't alive, although the cute little kitten asleep on it is. What distinction causes us to consider one to be living and one not? We only have direct knowledge of one form of life, life on Earth, and have only one data point from which to extrapolate theories about its chemistry and essence. Until life is encountered elsewhere, or aliens finally contact us, we will not have an independent second data set. Even then we may not, if the alien life itself shares an ancestor with life on Earth (we will come back to this idea later).

So What is Life?

To truly understand life, we need to explore what it is made of, what it looks like, and what it does. One of the oldest philosophical enigmas ever posed is that of life, and remains to this day a query with no (or quite possibly many) answers. Life on Earth is recognisable; we normally know a living organism when we see it as our brains identify a number of qualities all living organisms display. However, just because an entity presents one or all of these, there is the chance that it is still not technically a living organism. We cannot rely on our gut instinct and say something is alive because we *know* it is such – we need proof.

To start with, we can look at what life is made of. All life that we know about is formed of a close-knit family of units, popularly known as cells. Simple life forms are mostly made up of one single lowly but powerful cell (they are called *unicellular*), while advanced life forms, such as humans, are built of a magical mix of many millions and millions of cells working together (unsurprisingly, they are called *multicellular*). Cells do absolutely everything; they provide structure for the body, take in nutrients from food, convert those nutrients into energy, and carry out specialised functions that keep their host bodies working and running smoothly. They can do this because each individual cell is bound by a wall called a membrane that acts as a choosy barrier to the outside environment, sometimes letting in molecules and ions that the cell needs or pushing them out to keep the inside of the cell working properly. Cells are also the librarians of life. They contain the story of the organism they are a part of, holding its hereditary material within *deoxyribonucleic acid* or DNA and storing an instruction manual on how to re-build every aspect of that particular life form. Owing to life's dependence on the cell, materials in living organisms are always seen to exhibit some kind of unit order, and so we believe this order is a necessary condition for life to exist. Since cells and order break down over time, however, death for sure is also an inescapable characteristic of life, but alone does not make something alive in the first place. Besides, we hardly want to wait around for a potential life form to die, just to prove it was once living. As an example, this book has a structural unit order of paper and pages and this order can be broken and destroyed over time, but we would not consider it alive, or in fact dead.

For cells to do their jobs, and run the organisms they are a part of, they need energy and this is a common element always found flowing through living systems. Cells gobble up chemical compounds from the environment, transform them and spew out waste products as residues. In this way, cells obtain the basic materials to build their own body parts, and at the same time gain the energy needed to carry out the thousands of biochemical reactions that must happen every day, such as reproduction, growth, thought and movement. The need for a life form to get around is actually one of the most important uses of its energy. To enable it to survive, an organism must look for food, escape from predators and react to changes in the environment – if it starts to get too hot or cold for survival, then being able to move and find shelter or hunt for more comfortable conditions becomes really rather important. Sometimes an organism does not have to move physically but can interact with its surroundings and actually respond to the changes imposed on it. For example, when it gets hot humans and horses instinctively sweat, dogs pant, and elephants and rabbits send more blood flow to their ears to carry the heat away from their bodies. At a basic level, cells also respond to physical and chemical environmental stimuli and can communicate among themselves about how best to react through small signalling molecules. Using energy, enacting movement and responding to a changing environment are all activities that we have observed in living organisms, but those actions or responses cannot be used as universal indicators or markers to define life. A refrigerator utilises energy to regulate the temperature of its environment and a thermostat in your home increases the heat coming out of the radiators when it senses the room getting colder. Although we appreciate these objects and even in extreme cases rely on them for our own survival, they are most definitely not living beings.

The cellular custodians of life's instruction manual can be reproduced and in doing so pass on this hereditary information to the next generation of cells. Reproduction is when cells duplicate their genetic material and divide to produce two new 'daughter' cells, similar to their mother. All living organisms reproduce or at least are the products of reproduction. One exception to the former statement is mules, which are the hybrid products of reproduction between a horse and a donkey. Mules cannot reproduce themselves: they are sterile. Does this mean they are not alive? We remain reasonably certain that they are. Viruses are also tricky; they do reproduce, as the rapid spread of the common cold shows, but they cannot do it alone – they need to hijack the molecular machinery of a host they have infected. Does this mean they are alive or not? Defining life using an entity's ability to reproduce itself is a tough ask, as it means a computer virus could technically be considered a living organism.

Life grows and develops in patterns, yet the final forms, the offspring, are never a completely identical copy of the original.

This is the cornerstone of evolution – the passing on of heritable traits or characteristics, both good and bad, from parent to child and occurs within every generation born. This transfer of information to build a new being is, however, imperfect. In some circumstances, a mutant gene is introduced that imparts either a survival advantage or disadvantage for the newly created life form, making it better equipped to survive in the world it is entering, or possibly cause its untimely death. Evolutionary adaptation over generations is regarded as the most fundamental and unifying of the properties of life mentioned thus far. One of the most famous British scientists, Charles Darwin (1809–1882), has become synonymous with this fascinating skill of life. His research into what he termed 'natural selection' highlighted that those individuals in a population whose traits best enable them to survive and reproduce will go on and produce more offspring, which in turn will survive to reproduce. This is where the phrase 'survival of the fittest' came from. Evolution and an organism's chance of survival is essentially a genetic lottery.

There are many forces found in nature that fulfil nearly all of the above characteristics of life and could theoretically be classed as living organisms but we would still, in the face of all this proof, not say they are alive. Fire is a perfect example – it grows and moves as it encounters material to consume, providing it with more energy. As such it is self-sustaining, as long as the food remains available. It breathes oxygen and responds to changes in its environment. It excretes waste products of ash, heat and carbon dioxide by some of the same reactions that run the cells in our bodies, and it is even able to reproduce itself and make little fire babies. Like any life form, it can be brought into existence from a parent fire, or born from just striking a match. Even with all this evidence, we still do not say it is alive. Why? Is it because it hasn't demonstrated any intelligence or shown itself to have some kind of spirit or soul?

What if, instead of this list of physical characteristics that we can easily find exceptions to, we try and define life by what it does, rather than from its composition. I like this idea as it can incorporate life forms from the past, those being created now and those that will possibly be designed in the future. Who is to say that a robot or hologram that can think and make decisions, and may even be able to feel emotions or pain, is not alive, just because it is not made up of cells; for growth requires installation of a new part or algorithm instead of growing one itself; is born of a creator instead of a genetic parent; and does not store its blueprints within DNA or is silicon-not carbon-based? The question of what life really is remains by far one of the hardest, but also the most excitingly challenging to answer.

Carbon, Carbon Everywhere!

At the heart of all life is a single element: carbon. Every life form on Earth is built upon a skeleton of carbon and some also rely on a source of carbon-based food for energy. Since all atoms are essentially put together in the same way – with a nucleus housing a variable number of neutrons and protons, orbited by shells of electrons – what makes the

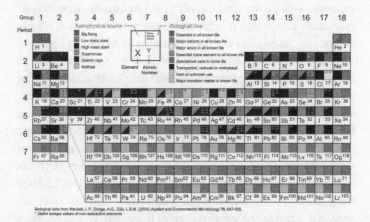

Figure 1 *The Astrobiological Periodic Table (credit: Charles S Cockell).*

carbon atom so special that life put all of its hopes for survival on it? The answer lies in the *periodic table*, an organised list of all the currently known elements in existence. Carbon has four electrons out of a possible eight in its outermost shell – it is half full. As the most stable configuration is to have all eight electrons present and accounted for, each atom of carbon has the ability to form up to four bonds with electrons orbiting nearby atoms to achieve this stability. The ability to form four bonds is not restricted to carbon – it's a property of every atom with four outer electrons, including those that sit below carbon on the periodic table, such as silicon, germanium, tin and lead. What's special about carbon is that it can form complex molecules built of double bonds, sharing more than one electron with other atoms, and these bonds are very strong. The simplest carbon molecules consist of a carbon skeleton bonded only to hydrogen atoms, unsurprisingly called *hydrocarbons*. There are more than a million known carbon compounds, broadly termed *organic* molecules, and it is finding these molecules that drives our current exploration of the Solar System and beyond. These organic building blocks are rife throughout the Galaxy, strengthening the idea of carbon-based life existing outside of the Earth.

The Units of Life

The cell mentioned earlier can be imagined as a membrane-bound miniature chemical reactor containing a library of genetic information, and is the building block of every living thing on Earth. The first known cells are thought to have originated in the oceans of the early Earth about 3.8 billion years ago. These were the *prokaryotes*, organisms made up of one single cell, such as bacteria and archaea. These cells do not have any internal organs, nor do they have a nucleus to house their hereditary material. Instead, their genetic instruction manuals float freely as a twisted closed loop of DNA within a watery cytoplasm inside the cell. For a billion years, the prokaryotes reigned supreme throughout the waters of the Earth. More or less 2.7 billion years ago, however, the Earth and evolution decided their dictatorship was over and introduced cyanobacteria to the world – tiny bacteria capable of converting energy from sunlight into food. They not only produced food but also became food themselves. The larger original prokaryotes found these tiny new bacteria highly appetising, and enveloped them in their plasma membranes. This resulted in a larger prokaryote with a smaller cyanobacterium inside; instead of the former digesting the latter, a symbiotic relationship developed whereby both organisms mutually benefitted from the new shared living situation. Over generations, these cyanobacteria stopped being a separate organism and became a part of the cell itself.

A second type of cell with internal organs (organelles), had arisen through evolution and this step changed the course of life on Earth forever. These cells, known as *eukaryotes*, have since been used to build everything from fungi to plants to humans. Most eukaryotic cells, otherwise known as animal cells, are invisible to the naked eye but can do everything from providing structure and stability to creating energy and a means of reproduction for an organism. We could not have evolved without eukaryotic cells, although interestingly in the average human body prokaryotic bacterial cells vastly outnumber eukaryotic human cells – so technically we are more bacterial than human. Eukaryotes evolved thanks to the predatory actions of prokaryotes gobbling up other prokaryotes who, instead of lunch, became a part of the original organism. However, even after their initial inception this eating of other organisms continued, and rather than being digested provided the eukaryotes with even more organelles, enabling them to evolve into ever more sophisticated cells. Eukaryotes are much more complex than prokaryotes and have a genome that is up to 10,000 times larger, housed in the control centre of the cell, the *nucleus*. Inside eukaryotes there are also mitochondria, which are the powerhouses of the cell, and perform reactions that extract energy from food. Specifically within plant and algae cells there are chloroplasts, which perform the light-harvesting reactions of a process called *photosynthesis*. Neither of these organelles is native to eukaryotic cells and chloroplasts are believed to have once been those free-living cyanobacteria that were engulfed by early prokaryotic cells. Over evolutionary timescales, this symbiotic relationship between the host cell and visiting bacteria developed to a point at which they were inseparable and fundamentally required each other for survival.

Food for Thought

Every living organism – from the smallest bacterium to the tallest tree and fastest mammal – needs a source of food and an input of energy flowing through its system to survive, driven by an intricate process of chemical reactions within each cell known as its *metabolism*. A strong metabolism allows organisms to grow and reproduce, digest and transport substances into and between different cells, maintain their structures, and respond to their environments. To power this metabolism, humans eat food, plants absorb sunlight, and microorganisms use energy produced from chemical reactions through a process called chemosynthesis. To make these reactions take place, a very special molecule called *ATP (adenosine triphosphate)* is used to transport the energy created by photosynthesis in plants, by cellular respiration in animals and by chemosynthesis in bacteria and archaea. As such, life cannot only be characterised by what it looks like, what it can do or whether or not its cells have a nucleus, but also by how it uses carbon and energy to metabolise. There are four categories that can be used to describe a life form based on the sources of carbon and energy available to them, and they are extremely helpful for hypothesising how and what type of life might be able to exist on other worlds.

Life can use carbon directly from the environment by drawing it from carbon dioxide found in the atmosphere or dissolved in water. This is how plants find carbon for their energy and so are classified as autotrophs. Other life forms derive their essential carbon from consuming pre-existing organic compounds through eating, such as animals and many microscopic organisms and these are called hetero-trophs. Human beings are heterotrophs. There are also two sources of energy available to life: plants use the power of sunlight through photosynthesis, termed *photo-*; and animals use chemical energy from reactions that happen when organic compounds are eaten, termed *chemo-*. Combining these carbon and energy sources we find *photoautotrophs* – organisms such as plants and microbes that absorb their energy from sunlight and obtain carbon from carbon dioxide in the environment. *Chemoautotrophs* draw their energy from chemical reactions using inorganic chemicals and derive carbon from environmental carbon dioxide. These organ-isms need neither food nor sunlight to survive and are found in environments where most other organisms would perish (we will meet a lot of these hardy little life forms later on). *Photoheterotrophs* get their energy from sunlight and carbon from consuming other organisms. This is rare but possible by some bacteria, such as *Chloroflexus*. Finally, *chemoheterotrophs* get both their energy and carbon from food – we humans are chemoheterotrophs. In general, eukaryotes can only feed on certain carbon sources, which can be pretty restrictive, but bacteria and archaea are truly remarkable as they can live off almost any imaginable foodstuff out there.

Molecules of Life

Life on Earth is built from only 24 of the greater than 100 known elements on our planet, each with properties that seem to be essential for a healthy metabolism. With so many elements available it almost seems a pity, and a tad risky, that just four elements – oxygen, carbon, hydrogen and nitrogen – dominate and control 96 per cent of the mass of a typical living cell. Life uses a polymer-based chemistry that includes nucleic acid polymers, DNA and RNA, to store and transmit information; carbohydrates for energy; and, when these are scarce or running low, fats. Yet the most diverse and multitasking of life's molecules are proteins, performing a vast array of functions within living organisms, including catalysing (speeding up or slowing down) metabolic reactions, replicating DNA, responding to stimuli, transporting molecules from one location to another and even building muscle. All proteins within living organisms on Earth are made up of the same 22 amino acids, from a selection of more than 500 found in nature. The key elements of every amino acid are the same as those of cells, namely carbon, hydrogen, oxygen and nitrogen. A fascinating quirk of amino acids is that there are two types; they are considered to be asymmetric molecules or *chiral*. To demonstrate this, hold out your hands in front of you. Human hands are perhaps the most universally recognised example of chirality: no matter how you try and orientate both hands it is impossible for all the major features of both hands to match up – a truly opposite mirror image. The same can be said for amino acids. There are, therefore, two possible forms characterised as either *left-handed* (L) or *right-handed* (D). Amino acids occur in both L- and D- chiral forms, but nearly all life on Earth uses the L-form. Sugars are also found in both chiral forms, although terrestrial living cells use the right-handed D-forms exclusively. Why life chooses these particular mirror images over the other is not yet understood. Organic material found in carbon-rich meteorites also seems to have a bias towards a similar handedness. The evidence is mounting that a predisposition to one form over another occurs naturally across all bodies in the Solar System and adds weight to the idea that the precursor molecules for early life are perhaps related to those in space or may even have come to Earth from the cosmos.

Could alien life have the same left-handed L-form amino acids as we do or would they use right-handed D-form instead? Perhaps they use left-handed L-form sugars instead of right? If all life in the Universe, both terrestrial and alien, spawned from the same pool of early molecules, then theoretically every molecule in the Universe would have the same chirality as is found on Earth. It would therefore be quite hard to tell if the life were truly alien, or just a very, very distant relative. The universe of chemical possibilities is huge. The number of different proteins that can be built from combinations of just the naturally occurring 22 amino acids is larger than all the number of atoms in the cosmos. Life on Earth certainly did not have time to sample and test all possible sequences and combinations to find the best. What exists in modern terrestrial life must therefore reflect some chance events in history that led to one choice winning over another, whether or not the choice was ideal. Perhaps some features of Earth's biochemistry emerged because of some now – unknown selective pressures on early life that no longer exist. Today's protein make-up and handedness may therefore not represent the finest design for survival in the modern world, but rather be a vestige of optimisation in an ancient one, such as is the case with the human appendix – once possibly needed in early humans for digesting leaves, today it is just a leftover of an ancient organ that has lost most or all of its original function. Who is to say, therefore, that life in the Universe may, or may not, have followed a similar pattern?

All life forms contain some form of DNA – the vessel containing all of the information-storing genes – life's genetic blueprint. If the entire DNA in just one of your cells were unpacked and stretched out straight, it would be nearly 2m (7ft) long. Since you have about five trillion (5,000,000,000,000) cells in your body and just over 2m (7ft) of DNA in every cell, the total length of DNA packed into our bodies would stretch from here to the Moon and back 1,500 times. It also has RNA (ribonucleic acid), which transfers information from the genes to enable the production of the cells' proteins. DNA is shaped like a long ladder twisted into a spiral – a double helix (RNA looks like one half of this ladder). Each strand of DNA's ladder has a carbon backbone of sugar molecules and phosphate groups and attached to it, making up the rungs, are chemical subunits known as *bases*. DNA is composed of four bases: Adenine (A), Thymine (T), Cytosine (C) and Guanine (G) and it's these letters that represent the code for building amino acids, that themselves make up proteins. The bases bind the two DNA strands together, with an A always bonding to a T on the opposite strand (and vice versa), and C and G doing likewise. A big question asked by biochemists is why DNA uses these precise bases, in particular adenine, when there are better alternatives? Maybe it was chosen over other available candidates by a freak accident, and was kept because later on it was thought too difficult to replace without losing fitness of the existing life forms. Potentially it was because adenine can be made prebiotically (chemically and before the formation of the first life forms) from ammonium cyanide, and had a much greater availability during the earliest eras in Earth history, making it a better choice in those times for starting life – even though a different contender might now be preferable.

No experiments can presently test all of the theories as to why life is built the way it is and uses certain molecules over others in its construction. But it's useful trying to understand the possible reasons for such choices, as it allows us to appreciate how easy alternative explanations are, and therefore the number of alien life forms imaginable.

Is Carbon Really the Only Option?

If for some now hidden reason life chose this amino acid over that one, left-handed over right, DNA over RNA, then what if it had not chosen carbon? For years, scientists and science-fiction writers alike have dreamed about the possibility of life based on some other element. To replace life's dependency on carbon would require a carefully chosen competitor. This challenger would have to be an element that is found in abundance across the known Universe and behave in a similar way to carbon, if life as we know it is still to function. *Silicon* is the first entertaining possibility. It sits nestled directly below carbon on the periodic table so has a similar personality. It has the same four electrons in its outer shell, meaning that it has four electron spaces available, giving it the ability to make four single bonds with other atoms, just as carbon does. It can bind readily to itself to make Si–Si bonds much like carbon can to other carbons, and it also bonds easily to hydrogen and oxygen given the right conditions. On Earth, silicon is more abundant than carbon. It bonded with two oxygen atoms and formed SiO_2 or quartz, the primary constituent of the rocks that make up the planet. The Earth is actually a silicon-rich, carbon-poor world with silicon unlikely ever to be in short supply.

So why on Earth did life choose carbon over silicon? One obvious answer is that outside the Earth and throughout the Universe there is much more carbon available than there is silicon, as fewer of the larger silicon atoms are formed within the cores of stars (we will explore how this happens in the next chapter). Silicon is used by life such as that found in the seashells abandoned along the beach but it is not the basis for any polymeric or metabolic chemistry. If complex silicon chemistry were possible on Earth, surely it ought to have resulted in life based on silicon, rather than its rarer chemical cousin, carbon. The answer may lie in the bonds that silicon makes with other elements, and how these may or may not be useful for life. For starters, it's a larger atom so the bonds it makes with other atoms under the conditions found on Earth are weaker than those made by carbon. There is also a huge difference between what happens when silicon and carbon bond with all the oxygen floating around the planet. Under the conditions found on the Earth, the molecule carbon dioxide (one carbon and two oxygen atoms) is a gas at most temperatures, is very soluble in water (and is therefore available within liquid solutions for life), and can be broken down into its constituent elements of carbon and oxygen – both of which are incredibly useful for life. In contrast, silicon dioxide (one silicon and two oxygen atoms) does not exist as a gas, except at extremely high temperatures over 2,000°C (3,632°F). As can probably be anticipated by the fact that it is the constituent of many rocks on Earth, silicon dioxide is almost completely immune to being dissolved; it's pretty solid. Finally, because silicon really loves to be bonded to oxygen, it is very difficult to break silicon dioxide into its constituent atoms. With respect to living organisms, silicon dioxide can be considered a very inert molecule and therefore somewhat useless for life processes. Consequently, carbon and carbon dioxide win the competition for being more useful to life, both as a molecule and split into individual elements.

Does this really matter when searching for alien life in the cosmos? Do the rules of chemistry work in the same way throughout the Universe? Would we observe silicon behaving differently on another planet if it had an environment unlike that of the Earth? Based on observations made by astronomers, the answer is probably no. Across the cosmic environment of the interstellar medium – interstellar clouds, meteorites, comets and stars – carbon molecules run rampant; not just simple ones, but also some of the more complex organic molecules as well. Oxidised silicon, such as silicon dioxide, is also quite commonly formed although silicon molecules such as silane and silicones that we would consider as silicon-based life molecules are seldom identified.

Perhaps counter-intuitive elements such as arsenic might be capable of supporting life under the right conditions? On Earth, some marine algae incorporate arsenic into complex organic molecules, such as arsenosugars and arsenobetaines. Several other small life forms use arsenic to generate energy and facilitate growth. It has even been speculated that the earliest life forms on Earth may have used arsenic in place of phosphorus in the structure of

DNA itself. Nonetheless, at no point has it been proposed as a possible replacement for carbon as the key to life. Titanium, aluminium, magnesium and iron are all more abundant in the rocks of the Earth's crust than carbon; so metal-oxide-based life could even be a possibility under some very non-Earth-like conditions found on a different rocky world. Boranes may also be an option. They are dangerously explosive in Earth's oxygen-rich atmosphere, but would be more stable in a reducing environment, one with little oxygen. However, boron's low cosmic abundance in comparison to carbon makes it rather unlikely as a base for life. What about chlorine and sulphur? Although purely hypothetical, sulphur could replace carbon, as it is capable of forming long-chain molecules just as carbon does. Some terrestrial bacteria have already been discovered to survive on sulphur rather than oxygen but have not as yet been found to replace carbon. Nitrogen and phosphorus could also potentially form biochemical molecules since phosphorus behaves like carbon in that it can form long-chain molecules on its own and, when combined with nitrogen, can create quite a wide range of useful molecules. Thus far, however, with no examples of any of these alternative life forms currently in existence, we only have one blueprint to build our assumptions from: that of carbon-based life. It seems, at least for now, that searching for life designed around carbon is the only truly sensible way to go.

Living is Thirsty Work

Carl Sagan famously dubbed Earth the 'pale blue dot' for its ubiquitous liquid. Water occurs naturally across the Earth's surface in all three phases – as a solid at the poles, a liquid in the oceans, and a gas in the atmosphere. Tasteless, odourless and virtually invisible as water vapour, it covers 70 per cent of our planet. The total liquid water on Earth is somewhere in the range of 1,260 million trillion litres

(326 million trillion gallons), although 97 per cent of this is undrinkable salt water filling the oceans and seas of the planet. Only two-and-a-half per cent of all the water on Earth is fresh water and all life living on dry land is reliant on this tiny percentage. Human life can use only a fraction of this, less than one per cent, and of that, about 70–90 per cent is used for agriculture. That does not sound like a great deal is left for us to drink, does it? But actually it is! Given the enduring presence of water on Earth's surface, it is not surprising that early life, and all subsequent life forms, were and are based upon and reliant on water. All life exists in an environment of water, whether it lives within it or uses it to form part of cell structures or as the main solvent in its metabolism.

It Came from Space ... Or did it?

Where this life-giving fluid came from, however, is hotly debated. The infant terrestrial planets were completely devoid of both water and carbon; they were simply too hot, what with being newly formed and recently molten. This means that the water required to allow life must have risen to the surface of the Earth somehow or come from somewhere. We know it showed up after the Earth's formation but probably only within the first billion years or so, either from deep within the cooling planet or from the reaches of space on board comets and water-rich meteoroids. Although the population of comets and asteroids passing through the inner Solar System is, thankfully for us, sparse today, it was a much busier time when the planets and Sun were young. Because our planet is in the Solar System's *Goldilocks Zone*, a region encircling the Sun where water has the opportunity to remain stable as a liquid, once the water molecules had surfaced they remained, and quite possibly played a key role in the development of life.

Until recently, it was believed that collisions with icy bodies from the outer Solar System likely brought much of the Earth's water. However, this theory was dealt a hard blow in 2014 as incredible results emerged from Europe's Rosetta mission (of which we will learn more in Chapter 5). This groundbreaking venture made history by landing on Comet 67P/Churyumov-Gerasimenko in November 2014, and revealed that the water on the icy body is unlike any found on our planet. While the vast majority of water on our planet is made up of hydrogen and oxygen atoms, very occasionally we find a hydrogen atom has been replaced with a deuterium atom. Deuterium is an *isotope* of hydrogen, and houses a single neutron in its nucleus, thus making it slightly heavier. On Earth, for every 10,000 water molecules, three deuterium atoms can be found. This water has the same physical properties, but owing to the addition of deuterium is heavier. Comet 67P was found to contain water that was 3 times heavier than water currently present on the Earth, which means that this variety of comet could not have brought water to our planet. This discovery adds to other studies that have analysed water on different types of comet, such as those that originated in the Oort Cloud – an icey region of space that makes up the outer reaches of our Solar System – which also has a different signature to water found on Earth.

Many scientists now believe that Earth may have had water from the start, inheriting it directly from the swirling nebula that gave birth to the Solar System. The conventional story followed the journey of carbonaceous chondrites (water-rich varieties of asteroid) that would have delivered water during the late stages of Earth's formation, possibly around 4.6 billion years ago, and these types of meteorites do provide some of the answers we are looking for. Carbonaceous chondrites have been dated as some of the oldest rocks in the Solar System, formed around the same time as the Sun, before the first planets, and they have isotopic signatures of hydrogen similar to Earth's seawater. However, it is now thought that the signatures of seawater have changed over geological time, gradually getting heavier.

The original seawater on Earth does not match that found within asteroids but has a hydrogen isotopic ratio closer to that of Jupiter and the solar wind. These are both thought to preserve the original isotopic signature of the solar nebula. As such, it is now thought that water may have snuck into our own growing planet, despite its scorching temperatures, by sticking to dust particles. Some of it may well also have arrived from space, although only around 10 per cent appears to have originated from comets from the Kuiper Belt and the Uranus–Neptune region of the Solar System. From the perspective of life, however, the source of the water is relatively unimportant – that it is there at all is what matters.

Although the exact mechanisms are poorly understood, it is clear that liquid water was present on the surface of the Earth only a short time after its formation. Excitingly, Earth is not unique in containing liquid water, however. Jupiter's moon Europa is covered with a sheet of ice that probably sits on top of a global salt-water ocean, and Saturn's moon Enceladus shows evidence of sub-surface water as well. Mars, meanwhile, was once a relatively wet world that apparently harboured large amounts of liquid water in the ancient past. It is the persistence of it in liquid form on the surface of the Earth that is unique and that we believe allowed for the gradual evolution of life.

A Special Liquid

The Earth is a wet and watery world so it should not come as much of a shock to hear that life makes good use of this abundant liquid. Despite its commonality, water is an extremely unusual molecule in its chemical and physical properties, and life has adapted to become entirely dependent on some of its unique characteristics.

Water as the molecule H_2O is made up of two hydrogen atoms attached to one oxygen atom. Water molecules are greatly attracted to each other and this stickiness is what gives water its high surface tension (imagine insects walking across a lake on what looks like a film). Water is the *universal solvent*, a powerful medium that surrounds and dissolves more substances than any other liquid currently known. Salts, sugars, acids, alkalis and some gases – especially oxygen and carbon dioxide – are hydrophilic (water-loving) substances. This is a very useful quality for biological processes. All of the components in cells (proteins, DNA and polysaccharides) are found within water, although not actually dissolved, instead deriving their structure and activity from their interactions with it. Other substances, however, are hydrophobic (water-fearing), such as fats and oils, and so are immiscible in water and will not mix, instead forming individual layers. Water allows substances in greater quantities to interact with each other at speeds faster than those obtainable in a solid, and slower than in a gas. Having molecules available in a dissolved liquid phase also helps cells to gether essential nutrients and expel waste products. Chemical reactions can take place in other phases as well, of course, but organic life is impossible as a solid or gas.

One of the incredible abilities of water is its response to changes in temperature. It remains liquid at a range of temperatures and pressures and is transparent in the visible electromagnetic spectrum. This matters because it allowed the rise of early photosynthetic bacterial and plant life, as sunlight was able to reach them through the overlying waters. Today, plants and bacteria have colonised bodies of water across the world, and in every environment. The boiling point of water (100°C (212°F)) and the freezing point (0°C (32°F)) is of utmost importance to the continuity and evolution of life, as throughout the last 3.7 billion years the temperature at the surface of the Earth has remained within this range at least somewhere on the planet. The range of places we can find life also benefits from using water as, as with all other liquids, it boils into steam at different temperatures depending on the air pressure. For example, at the top of Mount Everest water boils at 68°C (154.4°F), compared to 100°C (212°F) at sea level, regardless of latitude. Conversely, water deep in the ocean near geothermal vents can reach temperatures of hundreds of degrees and still remain liquid owing to the overlying pressures created by such a huge body of water. Most known pure substances become heavier as they cool; water, however, has the anomalous property of becoming lighter when it cools to form ice. It expands to occupy a nine per cent greater volume, which is why ice floats on liquid water, as evidenced by icebergs, and as an additional bonus insulates the water beneath from freezing. Water acts as a good temperature buffer as it can absorb a great deal of heat energy without a big rise in its own temperature. This skill benefits all life on Earth – on a global scale by helping the planet to steady its climate through stabilising the temperature of Earth's oceans, and at a cellular level by protecting an individual cell from wild temperature extremes that could disrupt and destroy metabolic enzymes.

Water is vital both as a solvent and as an essential part of many metabolic processes; it is fundamental both to photosynthesis in plants and respiration in animals. Photosynthetic cells use the Sun's energy to split off water's hydrogen from its oxygen. Hydrogen is combined with carbon dioxide (absorbed from air or water) to form glucose (energy) and releases the oxygen. Many living cells use such fuels and oxidise the hydrogen and carbon to capture the Sun's energy and reform water and carbon dioxide in the process (cellular respiration). Virtually every environment on Earth that has been examined seems to hold life that has evolved from the water-loving universal ancestor of all life on Earth. It seems that as long as long as water is available, life finds a way to exploit whatever thermodynamic disequilibrium exists.

What If There Were No Water?

Everything we know about life and its relationship with water suggests that Terran life (life on Earth) cannot exist without it. Yet, we can still ask the question as to whether water is specifically needed for life or if life is simply designed for a liquid environment and any form of solvent could be used? Perhaps terrestrial life evolved to exploit water simply because it was the only option to hand, so could life emerge in other, more widely available solvents on other worlds?

Ammonia, for example, shares many properties with water, and is actually quite analogous to it. It is liquid over a wide range of temperatures ($-78°C$ to $-33°C$/$-108.4°F$ to $-27.4°F$, at surface pressure on Earth) and an even greater range at higher pressures. It dissolves many organic compounds owing to the formation of hydrogen bonds, just as water does, and is abundant in the Solar System – it exists as liquid droplets in the clouds of Jupiter and within the dust of outer space. An ammonia or ammonia–water mixture stays liquid at much colder temperatures than plain water alone, so for the planetary bodies further away from their stars this could be a lucrative characteristic. Ammonia would not support the chemistry found in terrestrial life, however. Although alternate biochemistries could be formed and may one day be found right here in our own Solar System, perhaps on Saturn's largest moon Titan.

Another alternative to water is sulphuric acid. It's seen in the cloud layers above Venus and there are those who think life is possible, floating within these acidic aerosols. Perhaps Formamide is a solvent life might be able to use. Formed by the reaction of hydrogen cyanide and water, it is liquid across a wide range of temperatures, dissolves salts, and persists in a relatively dry environment, such as a desert. Hydrogen fluoride has also been proposed, as in theory it is a good solvent for both inorganics and organics vital to carbon–based life and has a larger liquidity range than water. The major difficulty is its extreme cosmic scarcity; but this is not a deal-breaker. Liquid hydrogen cyanide is another possibility and, unlike hydrogen fluoride, has a reasonably high cosmic abundance.

Now, if a high cosmic abundance of a solvent is an important factor for life, then the most abundant compound in the Solar System is surely worth considering: dihydrogen. It is the principal component (86 per cent) of the upper regions of the gas giants Jupiter, Saturn, Uranus and Neptune. So, is dihydrogen a liquid? Well, not exactly. Throughout most of the volume of gas giant planets where molecular dihydrogen is stable, it is a *supercritical* fluid – a substance that can effuse through solids like a gas (but isn't one) and dissolves materials like a liquid (but also isn't one). Little is known about the behaviour of organic molecules using supercritical dihydrogen as a solvent – one thing for certain is that the temperature at which dihydrogen goes supercritical is too high for stable organic molecules.

There is actually no need to focus strictly on polar solvents such as water when considering possible liquid habitats for life. Hydrocarbons such as methane, ethane, propane, butane, pentane and hexane are abundant throughout the Solar System and have boiling points up to 75.8°C (168.4°F) at standard pressures. Oceans of liquid ethane and methane have been observed to cover the surface of Titan. Perhaps if there were water droplets within hydrocarbon solvents on Titan, these bodies of liquid could be convenient cellular compartments for evolution. Indeed, pure hydrocarbon liquids may actually prove to be better than water for managing complex organic chemical reactivity. Methane could in theory support organic biochemistry although its low liquidity temperatures of −160°C (−256°F) may be too cold for biochemical reactions to run at the fast rates used by life as we know it to thrive. Perhaps life with slower metabolic processes could be possible?

All of these water replacements have pros and cons when considered in respect to our terrestrial environment. What needs to be considered is that with a radically different environment come radically different reactions. Life as we know it is built around a carbon scaffold using a water solvent; this has therefore become the standard chemical model we look for. Weird extreme environments may contain weird extreme life forms, so in time we may find that water and carbon are not needed to support life in the far-flung corners of the Solar System, although it is incredibly hard to imagine and design experiments to test for this today.

Everything that goes into creating a life form must come from somewhere. The rumour is we come from stardust – let's see how true that really is.

CHAPTER THREE

How to Create a Planet Fit for Life

What is it that allows for the formation of a gas giant versus a rocky moon, a comet versus an asteroid, a planet perfectly suited for the origins and evolution of life versus a barren, lifeless world? Life is resilient but also extremely fragile, so it needs to find a delicate balance of complementary conditions and some may say serendipitous events for it to originate, persist and thrive. To create a recipe for life, we need to cook up the perfect world for it to live in. The Earth is such a place, but its journey to becoming host to the only intelligent life in the Universe – as far as we know – has not been a smooth one, and several advantageous events have occurred to help it on its way. All life on Earth owes its existence to a single star that burst into light billions of years ago. In fact, it actually owes its gratitude to the star's death, not its birth. All the elements in existence, apart from hydrogen, helium and tiny amounts of lithium, descend from those cooked inside the fiery hearts of the first long-vanished stars – from the oxygen we breathe, to the carbon in our cells and the silicon in the rocks that built our planet.

A Cosmic Kitchen

Everyone knows the basic story of the creation of the Universe: there was a *Big Bang* and from nothing came everything. Okay, there is a little more to it than that, of course, but in terms of life, this multi-billion-year-long event can be summarised into a few key moments. The Universe is roughly 13.8 billion years old, a number hard to wrap the mind around with the sense of the hours we live every day. So instead of billions of years, let's imagine just one. If the entire creation of the cosmos were squashed into a single year, there would be 438 years per second, 1.58 million years per hour, and 37.8 million years per day – a cosmic calendar. In terms of the Big Bang, we need not bother with an entire year, just the first 15 minutes.

There might not have been an actual physical bang – especially if there was no one in existence to hear it go 'pop'. But this is as good a description as any of a process where one moment there was literally nothing in existence, and then suddenly the entire mass and energy of the Universe was ignited from a single extremely dense and hot spot called a singularity. At this moment, time officially started, the Universe began to expand and grow, the cosmic kitchen was open for business. There is no life without cells, no cells without carbon, no carbon without matter and all of this is created from three basic particles – protons, electrons and neutrons. A millionth of a second after the greatest cosmic event ever *not* witnessed, a single lowly proton was formed as the newborn Universe began to *cool* to a sweltering trillion °C (around 1.8 trillion °F). The proton's partner in crime, the electron, followed a second later, emerging from a broth of particles and antiparticles simmering at just 1 billion °C (a little over 1.8 billion °F).

After this initial birth by searing fire, the cosmos cooled further and grew increasingly menacing; the *dark ages* of the Universe had begun and lasted for the next 200 million years. During this time the Universe was a smooth soup of energy, its temperature hovering around 10,000°C (just over 18,000°F). The newly crafted protons, neutrons and electrons started to combine and the first atoms began to form. They initially assembled themselves into a single hydrogen atom, the most basic yet most abundant element in the Universe today. After hydrogen came helium and smatterings of lithium and beryllium. The basic ingredients for the recipe of life had been created but in this period the Universe was still completely sterile.

The newly created atoms effectively neutralised the Universe; it was no longer dominated by negatively or positively charged particles, which allowed matter to start to congeal owing to gravity, creating nodules or balls of concentrated matter that threaded across the still growing Universe. This is the birthplace of galaxies. Flurries of proto-dwarf galaxies formed – visually more akin to nebulae than the grand spiral and elliptical galaxies of today. These were the site for the creation of something without which life would not exist. They were the birthplace of the very first stars.

Star Light, Star Bright

Starlight came to the Universe some 200 million years after the Big Bang. These first illuminating orbs were made when a small parcel of gas within one of the newly birthed dwarf galaxies started to *feel* its own gravity and began a slow but accelerating inward collapse. The earliest stars were simple, formed only of hydrogen and helium as this was the only matter in the Universe to have been created so far. They grew within tremendously hot blobs of gas and were enormous. These first stars illuminated with an extraordinary brilliance but their life was fleeting, and they died in glorious supernova explosions that seeded the surrounding gas with their remains. Through these eruptions, nebulae of dust and gas were created that grew to become the stellar nurseries of the Universe. They provided the starting materials for each new brood of stars, which themselves died and flung even more elements, gas and dust into the Universe to be used to build generation after generation of stars – the stellar circle of life.

Without stars there would be no elements of the periodic table, and without elements there would be no Earth and no life. Stars are like giant nuclear reactors, constantly churning, creating and destroying elements. This deep nuclear fusion is what makes them shine. When their core reaches a high-enough temperature (a few million degrees) atoms are subject to tremendously violent collisions that release an enormous amount of energy; their nuclei begin to fuse together creating an entirely new element. In the early stages of a star's life, this reaction involves the nuclei of two hydrogen atoms combining to make deuterium, which after fusing with another proton, produces the light isotope of helium, He^3 - a process called *nucleosynthesis*. A star slowly converts its hydrogen into helium but there is not an everlasting supply. As the hydrogen fuelling the star's very existence is exhausted, nuclear reactions can no longer continue and the core begins to collapse under its own gravity. These higher temperatures cause the star to shine up to 10,000 times more brightly than before. The outer layers of the star then expand outwards, decrease in temperature, and the star becomes a *red giant* – technically more orange than red, but a giant nonetheless. What happens to the star after this bloating and blushing phase depends upon how large it is.

The smallest stars only convert hydrogen into helium, and that is the end of their life. Medium-sized stars (such as our Sun and those up to two times the mass of our Sun) will start to convert the newly formed helium atoms in the core into carbon and oxygen as temperatures reach 100 million degrees. Three helium-4 nuclei fuse together to create carbon, and then an addition of another helium to this carbon nuclei creates oxygen. The stars now have a core of carbon-oxygen. The largest or most *massive* stars (greater than five times the mass of the Sun) will convert hydrogen to helium, then helium to carbon and oxygen, followed by fusion of carbon and oxygen to form neon, sodium, magnesium, sulphur and silicon. Further reactions can take place, transforming the core into calcium, nickel, chromium, copper and finally iron.* Through all these stages of nucleosynthesis, each new *metallic* element created forms in the fiery soul of the star and is surrounded by a shell of the elements that came before it. If a slice were taken through one of these stars, it would display multiple layers of a giant elemental onion. This is what makes life possible and where the basic ingredients for life came from: stars. These blazing engines are the creators of carbon, the third most abundant element in the Universe, and by far the most important ingredient in our own creation.

But we have to ask – how do all these elements get from inside a star to our bodies? We mentioned before that the earliest stars that lit up the new Universe 200 million years after the Big Bang, exploded in what we call a supernova. The very low mass stars finish nucleosynthesis with a core of helium and a shell of hydrogen and have a quiet, dignified death. They do not go supernova. These stars simply start to become less luminous, ending their lives as a cool helium

---

* Most of the physical matter in the Universe is in the form of hydrogen and helium, so astronomers conveniently use the blanket term 'metal' to describe all other elements. So when the phrase *metal-rich star* pops up, the elements described are non-metals as far as chemistry is concerned, but considered metals in astrophysics.

white dwarf. When stars of great mass (8–25 times that of the Sun), however, have reached the stage at which they contain a fiery heart of iron, and no more reactions can take place, their elemental factory shuts down and the iron core collapses under the force of the star's gravity and implodes. This collapse releases a catastrophic wave of gravitational potential energy, causing an explosion that very briefly can outshine an entire galaxy – the star has gone supernova. This explosion spews nearly all of the star's matter and energy into space at up to 30,000km/s (around 18,640 miles per second, or 10 per cent the speed of light).

Supernovae are essential for the creation of life as they are a vital source of elements heavier than oxygen. Nuclear fusion within the cores of stars creates the elements lighter than iron and, although this is where the story ends for the star, it is not the end of nucleosynthesis. The power of the supernova explosion itself causes further chemical reactions that create even more elements, including plutonium and uranium. The Big Bang produced hydrogen, helium and lithium, while stars and supernovae synthesised the rest, enriching the interstellar medium and molecular clouds with metals.

What is left after a supernova is a compact object and a rapidly expanding shock wave of material heading into, and mixing with, the interstellar medium. This process, as brutal and destructive as it sounds, is actually good for life. These death throes of stars created and then seeded the CHNOPS elements (carbon, hydrogen, nitrogen, oxygen, phosphorus and sulphur) throughout the Universe – the elements that are needed to build life (and everything life needs).

The Gauntlet of Galaxies

Around 2.5 billion years after the Big Bang, and 2.3 billion years after the first stars sparked into life, gravity began to pull all the generations of stars thus far created into groups or clusters, commonly known as galaxies. A galaxy at its most basic level is a gravitationally bound system of stars and their remains – an interstellar medium of gas, dust and dark matter, all orbiting around a central point. There may be anything from a few hundred thousand stars to many hundreds of billions within a single galactic neighbourhood. Even within the small patch of Universe observable from the Earth there are hundreds of billions of galaxies and thanks to the work of famous astronomer, Edwin Hubble (1889–1953), we know that they will have one of four different shapes: spiral, elliptical, lenticular or irregular – but each as unique as a fingerprint.

One of the most familiar and intricately beautiful galaxy shapes is the spiral galaxy. In fact, when imagining a galaxy, this is what first comes to mind. This is because the Milky Way, the most famous of all galaxies, its neighbour Andromeda, and 77 per cent of all the galaxies so far seen, are winding, flat, disc-shaped spiral galaxies that loosely resemble an octopus. They basically consist of a central bulge (the head) with a number of different arms (the tentacles) spiralling outwards. Unlike those of an octopus, however, these arms are not restricted to eight in number. These twisted galaxies can be tightly wound coils of dust, gas and stars or loosely splayed tendrils, with all degrees in between. Since the Earth is currently orbiting within a spiral galaxy itself, it would be easy to think that spiral galaxies may be better suited to support life. Our continued existence in this rotating mass of stars and dust is the result of so many more factors than just its image, but the way its stars are able to move in well-defined orbits does make it a safer and more stable design of galaxy for the long-term prospects of life. The oldest observed spiral galaxy, BX442, is an impressive 10.7 billion years old yet coiled galaxies such as this are believed to be much younger than the less visually glamorous elliptical varieties.

Elliptical galaxies are the most massive with few or no dust lanes, being all central bulge and no disc. They are largely composed of older mature stars and seldom have stellar nurseries or new star forming regions. They have little to no rotation, so the stars display a variety of orbits, haphazardly moving around like a swarm of flies. As such stars commonly find themselves heading on a collision course with each other and the centre of the galaxy. As they draw closer, the increasing proximity and density of other stars produces an environment of high radiation and gravitational instability. If that were not chaotic enough, the statistical chance of these stars coming into close contact with more than one ancient star about to go supernova is much higher. These galaxies are therefore believed to be incredibly unfavourable for the emergence of worlds suitable for and capable of sustaining life.

A Galactic Goldilocks Zone

Planets qualified to support life are thought to be much more likely to exist around stars that reside in certain parts of a galaxy. The *galactic habitable zone* (*GHZ*) is a galaxy-wide Goldilocks zone, a theoretical ring threaded through a galaxy, where conditions exist that are favourable for supporting life should it happen to arise in any orbiting solar system. It covers a region lying in the plane of the Galactic disc that possesses enough of the heavy metallic elements that would be needed to build terrestrial planets like the Earth or Mars. As mentioned before, supernova explosions were responsible for the creation of interstellar dust clouds and these became increasingly more metal-rich with every additional stellar deposit, forming a nest filled with new baby stars. With more metal-rich material available, these younger stars were more likely to be able to grow families of planets to orbit around them. Galaxies, therefore, effectively have a *Goldilocks zone of metallicity*, a belt stretching across their waist whereby the amount of metals is just right to go into the formation of planets and where a planet fit for life can exist.

Within the GHZ, the cosmic environment needs to be sufficiently accommodating over several billion years to allow for the biological evolution of complex multicellular life, *i.e.* us. A major threat to this is the, up until now, very helpful supernova explosion. The blast waves created during the detonation, despite sending biologically useful elements into space, also release deadly cosmic rays, gamma rays and X-rays that can be fatal to any life form watching wide-eyed on a nearby planet or moon. This supernova fear factor is greatest in areas with the most stars and the largest amount of star formation. Keeping out of the way of the Galaxy's spiral arms is another requirement of a GHZ. The sheer density of gases and interstellar matter in the spiral arms leads to the birth of new stars. Although this is a good thing and can lead to the creation of planets and ultimately life, it would be dangerous for an already inhabited solar system to cross paths with one of them. The intense radiation and gravitational chaos of entering a spiral arm would cause catastrophic and life-threatening disruptions in our Solar System.

So to build an ideal planet suitable for life, we would start with a spiral galaxy and a metal-rich young star. This star would be orbiting around the core of the galaxy within a ring-shaped Goldilocks region at just the right distance from the galactic centre so that it has the minimum metallicity needed to form some rocky life-friendly planets, but is far enough away from the centre so that its solar system would not be continually plagued by swarms of exploding stars.

More Than Just Chocolate

After the chaos of the Big Bang, stellar births and deaths, and the formation of the first galaxies, there is really only one spiralling neighbourhood we are personally invested in: *the Milky Way*. It is home to the only example of life that we know of, and therefore rather an interesting place. It measures between 100,000 and 120,000 light years in diameter – a light year being a measure of distance, not time; it represents the distance that light can travel in a year, a cosmic speed limit, if you like. To put this expanse in perspective, light travels 9,460,528,398,225km (5,878,499,810,000 miles) over the course of one year, so multiply that by 100,000 years and the size of the galaxy is almost unfathomable.

The Milky Way is old, almost as old as the Universe itself. Recent estimates put the age of the Universe at 13.7 billion years, and our Milky Way has been around for up to 13.6 billion of those, give or take 800 million years. This is measured based on the age of the oldest stars in the our Galaxy, so the Galaxy must be older than they are. It is part of a larger family of at least 100 galaxy groups, each made up of 50 individual *Local Group* galaxies that include our neighbour Andromeda, forming a team known as the *Virgo Supercluster*. Within this family of galaxies, the Milky Way is moving through the Universe at a speed of 600km/s (over 370 miles per second). It contains between 200–400 billion stars but when you look up at the night sky with the naked eye, the most you can see from any one point on the globe is about 2,500. The number of stars in the Milky Way changes yearly owing to deaths and births; about seven new stars are born every year. Although this stellar headcount sounds impressive, the Galaxy is only a middleweight – the largest galaxy we know of is IC 1101 which has more than 100 trillion stars. Despite all this stellar illumination you cannot actually see 90 per cent of the Milky Way as most of its mass consists of dark matter that creates an invisible veil. In fact, every picture ever published of the Milky Way in its entirety is in fact not the Milky Way at all but another galaxy or an artist's interpretation. Currently, we cannot actually take a picture of the Milky Way from above because we are buried inside the galactic disc, about 28,000 light years from the galactic centre. It would be like trying to photograph your own house from the inside. But we do know that although incredibly beautiful and perfect for us, it is actually imperfect and warped. Neighbouring dwarf galaxies of the Large and Small Magellanic Clouds which are made up of a mere 10 billion stars are playing a game of tug-of-war with the Milky Way, pulling on its halo of dark matter and distorting the vast quantities of hydrogen gas. The result is a disc that resembles the profile of a sombrero.

The Milky Way is structured like billions of other spiral galaxies; it is not particularly special in that regard. Strong emissions of infrared radiation and X-rays leaking from its galactic centre have strongly hinted that clouds of ionised gas are rapidly moving around some sort of dark object – a black hole, and a supermassive one at that, called *Sagittarius A**, believed to measure about 22,530,816km across (14 million miles), or about the extent of Mercury's orbit around the Sun.

Around this black hole we find the highest density of stars in the galaxy and they are some of the oldest. Surrounding this 'nuclear bulge' is the galactic disc containing a lot of interstellar matter (dust and gas), as well as young and intermediate or middle-aged stars. Extending beyond this disc is a swollen 'galactic halo' where very old star populations are clustered. The high volume of stars in the central bulge influences the metallicity gradient spanning the Milky Way from highest in the galactic centre and decreasing outwards. Larger galaxies with a greater

Age (billions of years)

0   1.5   3   4.5   6   7.5   9   10.5   12

Figure 2 *Age map of the Milky Way (credit: M. Ness, G. Stinson/MPIA).*

number of stars tend to have a higher metallicity than their smaller counterparts. The Milky Way is by no means the largest galaxy out there, but it isn't the smallest either. Some 80 per cent of galaxies are less luminous than the Milky Way, but this is not vanity talking. A galaxy's brilliance is positively affected by its metallicity; the more metals present, the brighter a galaxy shines, and the more metals it has, the greater the likelihood of life arising. This statement depressingly puts 80 per cent of the Universe in a category in which life is far less probable.

The vast Milky Way halo and also the thick inner disc region are dominated by the older stars with low metallicity, so any planets that may arise around such stars are not predicted to have the materials needed for life. Some of the inner regions of the Galaxy, however, have the high metallicity required for the formation of terrestrial planets, but they are also unlikely to be suitable for life as they would be much more prone to suffering from extremes of radiation, being violently thrown around by gravitational fluctuations and hit by supernova shockwaves. The main region of the Milky Way that would be amenable for life is the thin disc where, coincidently, our Sun is found. The Sun's metallicity is used as a baseline (as we know it is good for life). Our Sun is exceptional in being both long-lived – currently 4.5 billion years young – and having a 40 per cent greater metallicity than most stars of the same age. Stars having between 60 per cent and 200 per cent of this level of metals are found in a region encircling the galactic centre at a distance between 15,000–38,000 light years – the Milky Way's GHZ. Unfortunately, for the possibility of other life-bearing worlds in the Milky Way, this area contains only about 20 per cent of the total stars in the Galaxy. Also, just because a star falls within this 20 per cent, does not automatically mean it has the potential to sustain life.

The Milky Way is not only moving but growing; it is a cannibalistic galaxy currently gobbling up hydrogen from those same Magellanic Clouds that are causing it to warp.

They are currently only 80,000 light years away from the centre of our Galaxy, technically already within the embrace of the Milky Way. They will probably be entirely absorbed into our Galaxy in about one billion years. The Andromeda Galaxy also seems set on a collision course with the Milky Way, a convergence that may begin about three billion years from now. The two galaxies will collide head-on and fly through one another, leaving gassy, starry trails, and scrambling stars to create new constellations. This may prove disastrous for any and all life forms, should they have remained in the area.

A Solar Family

The Solar System lies on the edge of the Orion–Cygnus spiral arm of the Milky Way, a nice secure distance away from the galactic centre, around halfway out, in a relatively uncrowded part of the Galaxy. Its orbit is remarkably circular around the galactic centre, which as a consequence keeps the Solar System safely away from the supermassive and super-destructive black hole, Sagittarius A*, at its core. A location in this suburban region also protects our solar family from the huge gravitational tug of stars clustered near the centre, keeping other planetary bodies such as planets, moons, asteroids and comets in their orbits and out of our way. The threat of any nearby stars going supernova and wiping us out is also reduced.

The Milky Way revolves around Sagittarius A* once every 250 million years but not uniformly, with the inner regions moving faster than the outer, and stars constantly overtaking each other. The four spiral arms do not rotate as a rigid structure either, so the orbit of the Solar System will inevitably pass through one of these spiral arms one day; every 100 million years to be exact, and it will take a further 10 million years to make it through, hopefully unscathed! As it takes this daring manoeuvre it will have to avoid increased supernova risks as the concentration of stars grows greater, and dodge the large dusty stellar nurseries, whose dust could infiltrate the Solar System and block out sunlight to any planets and moons. This has the potential to cause life reliant on sunshine to shut down and cool a planetary surface enough to cause an ice age. Luckily, however, the Solar System moves at nearly the same rate as the Galaxy's spiral-arm rotation. This synchronisation or 'co-rotation cycle' thankfully prevents our Solar System from crossing a spiral arm too often.

The Heart and Soul

The Sun is the source of all life on Earth and only its continued presence allows life to exist. A ball of gas held together by its own gravity, it is composed of hydrogen (92.1 per cent), helium (7.8 per cent), and less than 0.1 per cent metallic elements. The diameter of the Sun is about 109 times that of Earth (we could fit some 1.3 million Earths inside it) and is an almost perfect sphere, with a difference of just 10km (6.2 miles) in diameter between the poles and the equator. It is actually the closest thing to a perfect sphere that has ever been observed in nature. The Sun is a G-type main-sequence star (G2V), informally designated as a yellow dwarf, with a surface temperature of 5,500°C (9,932°F). The temperature in its core however is about 15 million °C (27 million °F) – it is a nuclear reactor after all.

About 4.6 billion years ago in a nebulous cloud of gas dust and stars far, far away, a grandparent star exploded in an incredible supernova event. It sent a powerful shockwave through the nebula, compressing regions sufficiently so that they began to collapse under their own gravity. The contracting gas started to move and rotate, whirling faster and faster, until it became a violently swirling storm that eventually broke apart into smaller vortices – each with the potential to form a solar system. The death throes of this single star gave birth to a whole brood of new stars and solar systems. One of these galactic tornadoes began to collapse and flatten into a rotating accretion disc swirling around a region that would become the point of origin for a single central star, the Sun. Of the entire disc, 99.86 per cent of its mass, which itself is 99 per cent gas, made its way into the Sun, but even the seemingly tiny amount of mass left was more than enough to make a family of planets, moons, asteroids and comets. The disc was enormous, stretching ten times the distance Pluto currently orbits from the Sun, and apart from gas was formed of tiny grains of dust – the seeds used to grow our planet.

The Rocky Worlds

As the Sun began to shine out across this foetal landscape, dust grains were drawn to each other, sticking and clumping together, growing into larger and larger lumps of rocky material. Once they became large enough, gravity caused them to be pulled towards each other. These rocky embryos became rounded because gravity pulls equally towards the centre of large masses, so anything jutting out was pulled back to form a ball. Within about one million years the accretion disc had become a rock garden, with several hundred rounded planetary pebbles orbiting around the infant Sun. Don't get me wrong, this was not a calm or smooth process. The Solar System at this time resembled a cosmic pinball machine with planetary embryos thrown in all directions by gravity, crashing into each other, breaking apart and recombining into larger bodies, or sent hurtling towards the Sun or even flung out of the Solar System entirely. Yet these colliding embryos became the building blocks for the rocky inner planets of the Solar System, including the Earth. The heat of these impacts and the decay of radioactive elements within the rocks themselves caused melting of the interior of the growing planets and the heavier elements such as iron sank to the centre, forming a core. These embryos took 100 million years to become fully grown planets but at this stage remained barren rocky worlds, lacking water, carbon and life.

The Gassy Worlds

A stream of charged particles (the solar wind) heading straight out from the Sun, swept away the lighter elements such as hydrogen and helium, from the Solar System's inner regions, leaving only heavy rocky materials to create the smaller terrestrial worlds such as Earth. Further away, the solar winds had much less strength, allowing hydrogen and helium to remain and be included into the building of the outermost worlds.

Just as when climbing a mountain an elevation is reached at which the ground begins to be covered in snow, at a certain point within the solar nebula an imaginary snow line forms, where falling temperatures freeze water and other volatile chemical compounds such as methane, carbon dioxide and carbon monoxide, and cluster them together. During the formation of our Solar System the water snow line was found around five times the distance from the Earth to the Sun, more or less where the asteroid belt is found today. Inside this boundary was a zone too warm for volatile compounds to be incorporated into the growing dust grains, so they became a gas and were lost. Beyond it, they condensed into solid ice grains and were built into the growing worlds. Not all chemical compounds have the same freezing point as water, however, so different molecules will freeze at different distances from a central star and may be the reason of why there are specific types of planet. For example, the carbon monoxide line in our system corresponds to the orbit of Neptune, and could also mark the starting point from where smaller icy bodies such as comets and dwarf planets like Pluto began to form.

The temperatures in the outer Solar nebula during this phase were well below $-120°C$ ($-184°F$), creating many more solid grains available for incorporation into planets – and so the large gas giants were born. Of the mass not taken up by the growth of the Sun, the remaining 90 per cent went into Jupiter and Saturn. The cores of these larger planets are rocky and icy but still more than 30 times bigger than the Earth.

This means that their gravitational attraction was strong enough to draw in large amounts of hydrogen and helium-rich nebula gas, making the planets even more massive, which in turn pulled in even more gas and dust – the ultimate snowball effect. At the furthest reaches of the Solar System there was little material left to make whole planets, so smaller icy bodies formed, like Pluto and the comets.

So there we have it – from a swirling nebula of gas and dust we have created a solar system family of inner rocky terrestrial planets and outer gas giants orbiting around a middle-aged yellow dwarf star. In actual fact, the Sun is encircled by eight planets, at least five dwarf planets, tens of thousands of asteroids, and up to three trillion comets and icy bodies. Yet, within this extended family there is only a single known world that supports one of the most precious commodities in existence: life.

Building a Home in the Goldilocks Zone

Planets, unlike stars, do not run a nuclear reactor in their cores and so are relatively cool, grateful for the warmth provided by their suns. In a star system such as ours, consisting of multiple planetary bodies, there is a special region that is uniquely suitable for planets to grow and nurture the conditions amenable to life; we call it the Goldilocks Zone, or more officially the circumstellar habitable zone or CHZ. In this imaginary region that encircles the Sun, surface temperatures on planets and moons have the potential to be able to sustain liquid water – they are wet. If worlds were to exist inside this zone and closer to the Sun, such as where Venus and Mercury reside, they would be too hot on their surface and any liquid water would boil away. They would be dry and desiccated. Planets and moons further out past the CHZ would be colder and any water present would freeze, as it has beneath the surface of Mars or on the surface of Europa. They would be frozen worlds (and incidentally dry worlds, too). Estimates of the CHZ boundaries in our Solar System are 0.75–0.95 astronomical units (AU) for the inner boundary and 1.37–1.90AU for the outermost, with the Earth positioned comfortably in the middle at 1AU (an AU being a unit of length roughly the distance between the Earth and the Sun). This positioning of the CHZ, however, is quite a simplistic view as for each individual solar system there are many variables to consider that might move or stretch its boundaries, such as the size and radiation of the central star, and the dimensions, mass and atmosphere-holding abilities of its individual planets and moons. Present CHZ models also do not include secondary regions where life not reliant on the Sun as a source of heat and energy could exist.

A Family Affair

Is it solely the distance a planet or moon sits from its sun that determines whether life arises or not? Although a critical dimension, there are many other factors, particularly in our Solar System, that have allowed Earth to hold this unique and privileged position. The first is, perhaps controversially, the protection Jupiter offers the Earth. The comet Shoemaker-Levy 9, discovered in 1993 by astronomers Carolyn and Eugene Shoemaker and David Levy, was observed in July 1994 from hundreds of observatories around the world as more than 20 fragments of it crashed into Jupiter's southern hemisphere. In July 2009, a second comet or asteroid this time ripped a Pacific Ocean-sized hole in its surface. This gives the impression that Jupiter is acting as a protective big brother – a celestial shield if you will, sacrificing itself and deflecting asteroids and comets away from the inner Solar System. The planet's enormous mass – more than 300 times that of the Earth – is enough to catapult comets out of the Solar System. It is also thought that Jupiter's gravitational pull could thin the crowd of dangerous asteroids and other objects, making Earth less impact-prone. Models of the formation of the Solar System suggest that the presence of a planet as massive as Jupiter also helped to conserve the Sun's angular momentum and stabilised the entire planetary system, especially the motions of the inner terrestrial planets. Three cheers for Jupiter ...

Next, it is perhaps the presence of our unusually large Moon that may have helped the Earth become stable enough to be a home for life. There is some dispute as to the origins of the Moon, but today the most commonly accepted theory is that about 30–50 million years after the initial formation of the Earth, a huge object with a similar mass and size to Mars smashed into it. As a result of this collision a significant part of the proto-Earth was ejected into space. This debris and that of the impactor coalesced into what became the Moon, one of the largest planetary satellites in the Solar System, with a diameter 25 per cent that of the Earth.

Earth's only satellite, it settled into orbit at a relatively close distance of about 30,000km (some 18,600 miles) but is gradually receding from us by about 3.8cm (1.5in) a year. Our Moon is also slowly *braking* the rotation of the Earth to the tune of about 1 second roughly every 67,000 years. These are both effects of tidal forces occurring between the Moon and the Earth. The Earth's rotation is slowing down due to rotational energy transfer to the Moon through the tides and because of this the Moon is very slowly increasing its orbital radius – and moving away from us.

The Moon has played a number of roles in the evolution and continued presence of life on Earth, although how necessary these have been is not entirely clear. The most familiar effect the Moon has is on the liquid envelope of the Earth, driving the tides, but should we lose it for some terrible (and unimaginable) reason, the Sun could in theory take over. Life wouldn't suddenly be extinguished because of this. The chief role the Moon plays pertains to the stabilisation of the Earth's axis over time. The tilt of the Earth is the main driver of the seasons, and this varies from 22.1 degrees to 24.5 degrees and back (known as the change in obliquity) over a span of 41,000 years, currently at a value of 23.4 degrees and decreasing. Without the large Moon to dampen this change in tilt, much wider and life-threateningly unpredictable swings would occur. This stability of Earth's seasons and climate has allowed for even the most complex multicellular organisms to evolve and thrive.

From the Inside Out

We know with some certainty the internal structure of the Earth – it looks a bit like the inside of a gigantic plum or peach. The Earth's innermost part, its inner core, is extremely dense and mostly made up of iron and nickel. It is unbelievably hot with a temperature of around 7,000°C (12,632°F) but instead of melting to form a liquid, it is completely solid. This is because of the tremendous pressures weighing on it, measured in gigapascals (GPa), as the mass of the entire planet pushes down exerting in the order of 360GPa. The source of these soaring temperatures this far underground cannot be due to the Sun. They are partly caused by left over heat produced during formation of the Earth and partly due to the decay of radioactive isotopes of potassium, uranium and thorium, whose half-lives are in excess of a billion years. This heat diffuses outwards, and makes a small contribution to the temperature balance of the Earth's surface. Interestingly, it also provides the heat and energy used by microbes that inhabit the outermost 3km- (1.86-mile-) thick layer of the planet.

The outer core is liquid and this 2,253km- (1,400-mile-) thick layer of iron and nickel moves or *convects*. This flow of metallic liquid is believed to create a geodynamo that influences the Earth's magnetic field, a shield that extends from the Earth's interior through the planet and several tens of thousands of kilometres out into space where it does battle with the solar wind. The magnetosphere deflects the Sun's charged particles and cosmic rays away from the Earth, protecting the atmosphere, which would otherwise get stripped away, thus protecting life on the Earth from exposure to deadly ultraviolet radiation.

Above the outer core lies the mantle, composed of hot but not quite solid rock. Its topmost layer, the asthenosphere, flows like a liquid but moves extremely slowly. This upper mantle layer and the outer crust together make up the lithosphere – the rigid outermost shell of the Earth – which is broken up into plates. The movements of these lithospheric plates over the mantle are known as the process of *plate tectonics*, and are the source of numerous phenomena that strongly affect life.

Earth's Recycling Plant

The seven major and many minor plates of the Earth's lithosphere ride atop the asthenosphere, travelling from only a few millimetres to up to 15cm (6in) a year. Each lithospheric plate can be topped by up to two types of crust – oceanic crust and continental crust. As their names suggest, one is created under the seas of the Earth and the other builds the land. The plates topped with continental crust can be up to 200km (124 miles) thick when they are carrying mountains, whereas those shifting oceanic crust are only 80–100km (50–62 miles) thick. There are no gaps between the plates; they all touch, forming a fractured but continuous rocky skin around the Earth. Where each plate meets another, however, their relative movements create different types of boundary: *divergent*, where the plates are slowly pulling away from each other; *convergent*, where the plates are colliding with each other; and *transform*, where the plates are slowly rumbling past one another in opposite directions.

As the lithospheric plates ride over the convecting mantle, new oceanic crust is formed along mid-ocean ridges, pushing the plates along and forcing the older oceanic crust back inside the Earth – a process called subduction. As such, these undersea rocks are relatively young (less than 180 million years old). Continental rocks, however, can be as old as 4 billion years – almost as old as the Earth itself. When two continental plates converge neither sinks and the plates buckle and crumple together, rising up to form massive mountain ranges. The Himalayas formed as a result of the Indian and Eurasian plates running into each other. Many of the characteristic features of the Earth, such as earthquakes, tsunamis, volcanoes, black smokers and mountain ranges, can be explained by these plate tectonics and the types of boundary between them – as such the consequences of plate movements are of great importance to the evolution and continuation of life.

The surface environment of the Earth shows long-term relative stability because by using plate tectonics it can regulate its own temperature through a process called *the carbon cycle*. It works like this: carbon dioxide from deep within the Earth is constantly pumped into the atmosphere via volcanoes and deep-sea hydrothermal vents, which in turn are made possible by plate tectonics. But the carbon dioxide from volcanoes does not stay in the atmosphere indefinitely. It is actively removed by chemical weathering whereby it reacts and combines with the rocks on the Earth's surface, and is then returned to the inside of the planet through plate subduction, preventing any life-threatening accumulations of carbon dioxide from occurring in the atmosphere. This process is temperature sensitive and works faster at higher temperatures. By this mechanism, the Earth regulates its temperature and keeps it habitable for life: if the planet's atmosphere starts to get too hot, the weathering rate will increase and more carbon dioxide will be drawn down and chemically captured within the rocks of the Earth. The concentration in the atmosphere will then decrease, reducing the greenhouse warming caused by too much carbon dioxide, and the temperature of the planet will drop. It is a delicate balance though. If the temperature of the planet drops too much and it starts to freeze, tectonic processes will work to ensure that carbon dioxide is pumped back into the atmosphere to help it warm up again. But without liquid water (as it would all be frozen) there would be no carbon dioxide removal by weathering, so the carbon dioxide concentration would build up in the atmosphere until the temperature rises to the point at which the ice melts, and weathering commences again. During weathering, carbon dioxide is converted to a soluble ion known as bicarbonate (HCO_3^-), which precipitates in the oceans as minerals such as calcite and dolomite that go into making seashells, coral reefs and the white cliffs of Dover. These minerals are decomposed when subducted and drawn back into the Earth, releasing their load of carbon dioxide into the mantle, ready to be erupted again by volcanoes. The cycle is complete and the Earth, through plate tectonics, maintains a warm, water-rich environment suitable for both simple and advanced life.

What would happen to life if the tectonic plates stopped drifting? Well, the Earth would be a very different place. The volcanoes of the Pacific Ring of Fire, in South and North America, Japan, the Philippines and New Zealand, for example, would shut down, and there would be far fewer earthquakes. Erosion would continue to wear down the mountains, but with no tectonic activity to refresh them, over a few million years the whole planet would be a great deal flatter. The level of the seas would rise as the polar caps melted, and most of today's dry land would be submerged. Only a small number of isolated dry islands would survive. Would intelligent life have arisen on Earth if it were a mostly flat aquatic planet? We don't know. If the plates stopped moving, the planet would need to find a new way to regulate its temperature if life were to survive. It is not clear what that mechanism might be, or even if one exists. Perhaps the Earth's crust would appear like the single plate crust of Venus and fall victim to catastrophic volcanic episodes? What is clear to us is that for a world to be habitable for life it needs a way to regulate its temperature so as to keep it suitable for its indigenous life forms. As far as we know, plate tectonics is the perfect mechanism to achieve this.

Lucky Quirks

Part of the handicap we face when designing the perfect world for life to thrive is that there is only one planet in the Solar System where we can currently observe processes such as plate tectonics – any evidence for it on Venus and Mars is at best very tenuous. Life on Earth is adapted to the effects of plate tectonics, but until we find another example somewhere nobody can say if tectonics are crucial for life to exist. There are, however, three factors that we know with complete certainty are essential for life: carbon to build cells, water as described in Chapter 2, and energy; and the availability of each of these is linked to special properties of planet Earth.

The where and why of how we think the Earth obtained its water has also been discussed in Chapter 2, but the most important aspect for life is the fact that we have it and it is wet. Most of the planet is at the perfect temperature, fluctuating within the boiling and freezing points of water, so for the most part water stays liquid. Apart from the role plate tectonics plays in this, where the Earth sits in our Solar System, in the Goldilocks Zone, is the main cause. The amount of solar warmth that envelops the Earth is dependent upon the Suns' brightness as dictated by its dimensions and chemical composition. All planets and moons in the Solar System receive some degree of warmth from the Sun, but their distance from it is key to whether they receive too little, too much, or just enough. As luck would have it, the Earth sits at the perfect distance, where the warmth it receives is *just right* to allow water to be a liquid on its surface.

Distance from the Sun is only one lucky factor that makes our planet nice and cosy for life. The Earth also has an atmosphere full of greenhouse gases, in particular carbon dioxide, which help to warm the surface of the planet below. Without any greenhouse gases and their warming effects, and with the Earth's surface reflecting sunlight away back into space (an effect called its albedo), Earth would be frozen and hover around $-15°C$ ($5°F$). It is not as simple as Earth sitting at the right distance from the Sun therefore: the Earth *itself* promotes life. It can be easily imagined that higher or lower carbon dioxide levels would be necessary to maintain a habitat on a planet whose distance from its Sun is lesser or greater, in response to the volumes of solar radiation it receives. The outer reaches of a habitable Goldilocks zone are achieved when the levels of carbon dioxide in an atmosphere become so high that it forms clouds, blocking incoming solar radiation from reaching the surface and causing an increase in the planetary albedo ... the end result is a frozen world.

The radiation-absorbing talent of Earth's atmosphere also supports this ideal surface temperature for water and consequently life. Our atmosphere has a window; it allows some infrared radiation from the cloud tops and surface to pass through it directly to space without intermediate absorption and re-emission, and thus without heating the atmosphere. Without this infrared atmospheric window, the Earth would become much too warm to support life, and possibly so warm that it would lose its water and come to resemble the planetary greenhouse that is Venus.

A third atmospheric quirk is that it is thick enough to exert a pressure on the surface of the Earth, which suppresses the rapid evaporation of liquid water. Atmospheric pressure is the force per unit area exerted on a surface by the weight of air above. On Earth, the atmosphere around us is filled with air molecules that collectively weigh on our bodies. Although you cannot feel it, Earth's atmosphere presses down with the force of $1kg/cm^2$ (or some $14.7lb/in^2$) and terrestrial biology has evolved to operate quite easily under it.

The final special property of the Earth is its gravity. Gravity allows the planet to hold on to its all important atmosphere, allowing only a little to escape into space, which is quickly replaced by outgassing volcanoes. At the same time, the Earth's gravity is not so strong as to attract a denser atmosphere, which would over-insulate the surface and produce increased surface temperatures unacceptable to life.

The Earth has proven itself physically fit and highly adaptable which enables it to support and nurture life. Luckily it arose within the Goldilocks Zone of the Solar System, which itself formed and sits nicely within the Goldilocks Zone of the Milky Way. The next step is for life to arise and take advantage of this perfectly situated and uniquely designed world.

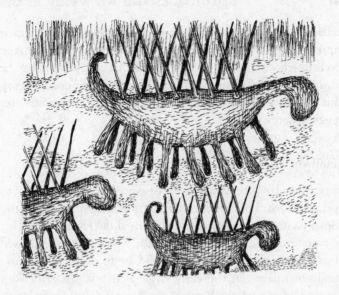

CHAPTER FOUR
The Story of Life

Once a planet has been prepped for life, the time hopefully comes to populate it. Of all the stages that happened during the Big Bang and beyond, stellar nucleosynthesis is by far the most important for this. Remember that nuclear fusion within the cores of the early stars created the heavy elements, such as the biologically useful carbon, nitrogen, oxygen, phosphorus and sulphur, with iron the end of the road. The outer layers then detonated violently as a supernova, hurling all these biologically crucial elements out into the cosmos. These bonded over millennia thanks to UV radiation to create simple compounds such as water, formaldehyde, ammonia, hydrogen cyanide and hydroxyl. Astronomers have so far identified 130 different organic carbon-based molecules in space, the most common being polycyclic aromatic hydrocarbons (PAHs) – precursors that are central to the development of life on Earth. The Galaxy is very encouraging for life; the building blocks of terrestrial biochemistry are everywhere.

Hell on Earth

Let us go back to when time began on the newly created planet Earth. The first stretch, starting from the instant the Solar System began to form, is termed *the Hadean eon*. This covers the events during which the Earth metamorphosed from a gaseous cloud into a solid body of rock. Because collisions between the early large planetesimals released a great deal of heat, the Earth and other newly formed planets would have been molten, only starting to harden as they cooled down. The traditional view of conditions on the Earth during this time is what led to the time period's name: Hadean, from Hades, the Greek mythological underworld. It was seen as a steaming, lava-filled 'Hellish' period in history. If we were able to travel back to visit the Earth at that time, it would probably not remotely resemble the planet we know and love today. Opinions about what it was really like on Earth, especially for life, are mixed, as there is very little evidence to work with. But what *do* we know?

In the beginning the Earth was an almost perfect sphere of molten rock, a burning landscape pelted by rocky leftovers from the formation of the Solar System. The common perception is that the young Earth was a hot, deserted world peppered with pools of simmering magma and with an environment that was inhospitable for life. Impacts caused the Earth's surface to be submerged again and again under large volumes of lava – enough to cover the globe several times over in a molten layer of scum. The heavy elements, such as iron, began to sink through this gloopy shell towards the centre of the Earth while the lighter ones, particularly the silica-rich minerals, formed an incandescent ocean covering the surface. Approximately 500 million years after the birth of the Earth, this sweltering panorama started to cool off and rocks began to form on its surface in regions that were in contact with the cold surrounding envelope of space. However, this delicate rind was forced to melt and re-form numerous times as gigantic magma currents erupted from the depths of the planet, while colossal rocks from space came soaring in to tear the new crust apart. Once evidence for hard rock forming on the surface was observed, the geological history of the Earth officially started; the Hadean eon ended and the *Archean* one began.

We recognise that life needs an atmosphere to allow it to arise and take hold on the surface of a newly formed planet, but during the earliest millennia on Earth, even though it did have a primordial atmosphere, it was very different to that which surrounds the Earth in our times. It was probably a reducing atmosphere, meaning it was lacking oxygen, and would have been toxic to nearly all life that exists on the planet today. The combination of exceptionally high temperatures and extreme volcanic outgassing of water, methane, ammonia, hydrogen, nitrogen and carbon dioxide had created this atmosphere. Interestingly, early Earth's atmosphere is quite similar to the current atmosphere of Saturn's moon Titan. The primitive Earth was wrapped inside a blanket of dense burning clouds and remained shrouded in darkness. When temperatures finally cooled sufficiently, the clouds began to drip; the first water droplets started to rain down and the Earth was assaulted by extreme weather events of enormous proportions. At first falling on flaming rock, the rain instantly evaporated, but over time it gradually cooled the crust enough to enable water to collect in the depressed regions of the Earth's surface, forming the first oceans.

Given this picture of a tumultuous and dangerous infant planet, if some hardy organism had somehow popped into existence, surely it would quickly have been extinguished, perhaps by one of the giant meteorites that slammed into the Earth. However, computer simulations conducted in 2014 suggest that the early Earth may not have been as hellish as was previously thought. The common thinking until recently was that life could not have emerged on Earth until the bombardment of projectiles from space eased and the surface was able to solidify to some degree. However, it is now thought possible that between these impacts there were tranquil times when oases of water could have existed and even have supported the early evolution of life. It is not yet known whether life emerged and was then snuffed out by each later impact, or if it never had the chance to take hold in the first place.

Over the last decade, small hardy crystals known as zircons have been found embedded in ancient – we are talking billions of years old – Australian rocks, and have painted a picture of the Hadean period completely inconsistent with the myth. Zircons up to 4.4 billion years old suggest there was liquid water on the surface of the Earth soon after it formed and that plate tectonics had already started. Analysis of the relative amounts of different isotopes of oxygen inside the crystals show that the ratio was skewed toward *heavy* oxygen-18, as opposed to the more common *light* oxygen-16. When a geologist sees a heavy oxygen signature in rocks, it is commonly understood to be a sign that the rocks formed in cool, wet, sedimentary processes at the Earth's surface.

We now almost universally agree that by at least 4.2 billion years ago, the Earth was actually a reasonably placid place with land, oceans and an atmosphere – representing relatively suitable conditions for the origin of life. An understanding of how, why and where life first arose, however, still mostly eludes us.

It's Raining Rocks

Whether or not you believe that the early planet was a real-life Hell on Earth, it nevertheless was a treacherous place for life to arise and be sustained, and this life would have had to be extremely resilient and incredibly tough (we will explore what this life might have looked like in Chapter 6). Sadly, however, any geological evidence that could help us solve this mystery is missing. The oldest meteorites and lunar rocks are about 4.5 billion years old, whereas the oldest Earth rocks currently found are only 3.8 billion years old. Why? During the first billion years after its formation, the inner Solar System was crowded with debris and the newly born planets underwent a lengthy bout of cosmic bumper cars with comets and asteroids. Astronomers believe that about 600 million years after the Solar System was formed (or some 4 billion years ago), a vast expanse of space beyond the orbit of Neptune, the Kuiper Belt, was shaken up by the migration of the gas giants Jupiter and Saturn. This gravitational disruption scattered comets and other icy bodies, flinging many into interstellar space but also throwing some on to orbital paths that wreaked havoc on the inner planets of the Solar System. This period is affectionately known as the Late Heavy Bombardment and lasted hundreds of millions of years. Around 3.85 billion years ago this cosmic assault of the Earth finally ceased and the surface was able to solidify. These impacts, along with erosion and plate tectonics, destroyed or buried nearly all of the rocks older than 3.8 billion years, concealing this period of time from us forever. Although no rocks from that time exist on Earth today, we have another source of information – the Moon. Many of the numerous craters and lava flows decorating the Moon's pristine surface provide a record of the Late Heavy Bombardment, and judging by their diameters of hundreds of kilometres, this was a violent and destructive period in the history of the Solar System.

An upside to the Earth being blasted by space rocks (yes, there is one) was that this blizzard of comets and asteroids from beyond the Snow Line delivered enormous amounts of crucial volatiles to the Earth's surface, such as additional water, carbon dioxide and simple organic molecules.

We are Aliens

Could the carbon within these asteroids, comets and the dust arriving on the early Earth be the same carbon used to kick-start life? It is easy to imagine these cosmic deliveries of prebiotic organic compounds having played a part in the story. Hundreds of tonnes of organic carbon are still delivered to Earth every year, and the rate could only have been higher in the chaotic young Solar System. Today, carbon is transported to the Earth inside meteorites – such as within the carbonaceous chondrite Murchison, named after Murchison, Victoria, Australia where it was seen falling from the sky in 1969. It is one of the most studied meteorites in the world, simply because there is so much of it: more than 100kg (220lb) of space rock. The story goes that on 28th September at 10.58p.m. a bright fireball was observed to separate into three fragments before disappearing in a cloud of smoke. About 30 seconds later, a tremor was felt. Many fragments were found over an area larger than 13km^2 (5 square miles), with individual fragments weighing up to 7kg (15lb); one even broke through a roof, falling into a pile of hay. Regardless of this obvious terrestrial contamination upon landing (even though the pieces were quickly found and collected), these meteorites are extremely carbon-rich. They carry the signature of the Solar System from the time the Sun was born 4.6 billion years ago, freezing in time snippets of billion-year-old chemistry. Within this single rock the diversity of prebiotic organic molecules is truly staggering. It contains amino acids, carboxylic acids, polycyclic aromatic hydrocarbons (PAHs), nucleobases, alcohols, aldehydes and ketones. Excitingly, over 70 amino acids have been found, even though life on Earth only uses 20 and only six of these 20 needed for life were found in Murchison. The rest are completely alien to life on Earth.

Some comets may have transported water to the early Earth but they also brought organic compounds. These dirty snowballs, or in fact snowy dirtballs, are leftovers from the dawn of the Solar System and contain dust, ice, carbon dioxide, ammonia, methane and much besides, preserving evidence of chemical processes that were at work billions of years ago. The nuclei of most comets, which are coated by a dark layer of organic material, are thought to measure up to 16km (10 miles) across. This represents an enormous store of carbon-rich goodness. An early result from the Philae Lander's first suite of scientific observations of Comet 67P/Churyumov-Gerasimenko in 2014 revealed that it supported 16 carbon- and nitrogen-rich compounds. The significance is that some of these compounds play a key role in the prebiotic synthesis of amino acids, sugars and nucleobases: the ingredients for life. Carbon within organic molecules is also delivered to Earth within Interplanetary Dust Particles (IDPs). These are extraterrestrial grains fondly called *cosmic dust*, which have been collected in the stratosphere by high-altitude aircraft. These particles comprise different minerals, mainly silicates and, importantly, a carbon-rich material containing hydrocarbons (CH_2 and CH_3) and carbonyl ($C=O$) that are used by life.

Scientists have long debated the possibility that the seeds of life did not originate on Earth. However, instead of the deliverance of prebiotic compounds onboard space rocks, it has also been suggested that microbial life may have travelled here fully grown from Mars or even another star system, and then evolved into the plethora of species seen today. This idea is called panspermia, a highly controversial concept that microbial life is everywhere in the Universe and can spread between planets on board comets, meteorites and dust. In essence this theory suggests we may all be Martians, or even Europans. Although an explanation favoured by few scientists for the origins of life on Earth, there are aspects to it that are intriguing. To get here, simple life forms would have had to endure a litany of harsh cosmic conditions, including ejection into space from their home world on board a rock; the freezing temperatures, radiation and vacuum of space; the million-year timescales involved with the journey to Earth; a fiery re-entry through our thick atmosphere; and finally a high-speed impact into the solid rocky crust. It is proposed that as long as any organism is buried deep enough within a rock of reasonable size and is able to remain in a dormant state over geological time, it might be able to survive the ride to Earth. To transfer a rock between Mars and Earth could take up to 15 million years, since it is necessary to wait for its orbit to cross that of the Earth. That is an extremely long time for life to remain dormant and to survive, and currently we have no idea if it is possible. We know that planets and moons have exchanged rocks before, as evidenced by 132 meteorites arriving on Earth from Mars and 180 from the Moon, and from the photos of meteorites sitting within the surface dust on Mars. There is therefore a chance, albeit slim, that life rode in on one of these rocks and made itself a new home.

From Soup to Cells

Aside from microbes riding in upon a meteoritic chariot, what are the possible routes that life might have taken to arise on the Earth? Where might this miraculous event have taken place? And most importantly ... how quickly after the planet had coalesced from primordial dust and gas did chemicals manage to organise themselves into life? Some astrobiologists approach it from the present, moving backwards in time from complex multicellular life today to its simpler unicellular ancestors. Others march forwards from the formation of Earth 4.55 billion years ago, exploring how lifeless chemicals might have built living beings.

How? Where?

Let's admit this up front – we do not know exactly how life got started, but we do know that all life on Earth is related. Living things (even ancient supposedly *simple* organisms such as bacteria and archaea) are enormously complex. However, all this complexity did not leap fully formed on to the Earth's surface from just a combination of a few simple elements. Instead, life almost certainly originated in a series of small steps, each building upon the complexity that evolved from the last. Humans and chimpanzees share a common ancestor from at least 7 million years ago; humans are related to the first mammal that lived some 220 million years ago, and together with bacteria have evolved from a shared family member who lived billions of years ago. The oldest evidence of life on Earth turns up about 3.9 billion years ago and has possibly been found in rocks formed 4.2 billion years ago. But what was there before that?

We suspect that ancient organisms shared the same basic traits found in all free-living organisms today – encoding genetic information in DNA and running a metabolism via proteins. DNA and proteins, however, are a paradox – they depend on one another for their survival – so it is hard to imagine one of them having evolved first without the other; although it is just as implausible for them to have emerged together. Chicken and egg! We now think earlier forms of life may have been based on a third kind of molecule found in today's organisms: RNA. Overlooked for many years, RNA turns out to be astonishingly versatile, not only encoding genetic information like DNA but also acting like a protein, carrying out the functions required to keep a primitive cell alive. This *RNA world* may have spurred life into being, although hardier molecules were required to take it further. Once proteins emerged, they would have been favoured by natural selection, as they are thousands of times more efficient as a catalyst. Likewise, genetic information can be replicated from DNA with far fewer errors than it can from RNA.

Just where on Earth these building blocks came together as primitive life forms is a subject of debate. Life started in water – this is probably the only aspect universally agreed upon. In 1871, Charles Darwin speculated that it may have begun in a 'warm little pond' and in the 1920s this became known as the *primordial soup*. Based on a theory of a chemically reducing atmosphere and energy from intense episodes of lightning, simple organic compounds may have been created in the atmosphere. The theory suggests that these rained down on to the Earth and accumulated in a liquid pool within which further transformations occurred, creating more complex organic compounds and ultimately life. This idea has fallen by the wayside somewhat, not least because it has been realised that Earth's atmosphere would not have been as reducing as previously thought, owing to the immense volumes of carbon dioxide being pumped into it by volcanoes, and this makes the production of organic molecules in this fashion slightly more challenging, although not impossible.

Starting in the 1980s, many scientists argued that life started in the scalding, mineral-rich waters streaming out of deep-sea hydrothermal vents (fissures in the Earth's surface from which geothermally heated water spews). Here there would have been heat, chemical energy and minerals such as pyrite, or clays that would have provided reactive surfaces to stabilise the organic molecular building blocks of life. Evidence for a hot start included studies on the tree of life, which suggested that the most recent common ancestor of all life seen today was an aquatic microorganism that lived in extremely high temperatures – a reasonably good candidate for the inhabitant of a hydrothermal vent! Nowadays, the hot-start hypothesis has cooled off a bit. If life did appear at hydrothermal vents, the temperatures would have needed to be below 80°C (176°F), or organic macromolecules would not have been able to survive.

Over the last few years, a slightly different picture has emerged of life beginning inside warm, gentle springs on the sea floor that bubbled billions of years ago when Earth's oceans took over the whole planet. These springs – as opposed to the scalding hot acidic hydrothermal vents – would have been cooler and alkaline. Early Earth's oceans were rich in carbon dioxide, as was its atmosphere. When carbon dioxide from the ocean met with hydrogen and methane from the springs over the chimney wall of a vent, electrons may have been transferred, producing reactions that created more complex carbon-containing compounds – essential ingredients for life as we know it.

What is most likely, however, is that life did not kick off from a single spot on the Earth at a single moment, but that the very early cells appeared multiple times in multiple localities. The successful early cells would have colonised all available habitable sites, transported by ocean currents. It is quite possible that the first organic substances arose from a combination of sources as well – from reactions in the atmosphere, rocky reaction chambers on the ocean floor, and even via delivery from space. Over time, life would have run out of available resources in its local environment and have had to adapt to take advantage of other potentially habitable locations, or face extinction. The investigation continues ...

Chemistry Becomes Biology

We still do not know at what point we might consider an early organic molecule to have been *alive* and how this might have happened, which presents us with just one more 4-billion-year-old mystery to solve. We can, however, speculate as to what this first living microorganism might have looked like and how it might have gone about its day. This first microbe is commonly called the last universal common ancestor (LUCA) as it is the most recent common ancestor of all current life on Earth. Unfortunately, it would have been far too fragile to be preserved within the fossil record for us to find today (or at least no evidence of it has been dug up so far). Nonetheless, we have a few ideas about what it might have been like. It would have been a small, single-cell organism with a cell wall and a freely floating ring-shaped coil of DNA – a little similar to modern bacteria. While its appearance and anatomy are slightly uncertain, the internal mechanisms can be understood based on the properties currently shared by all independently living organisms on Earth. For example, all life today uses a DNA/RNA genetic system and proteins to power its metabolism so the LUCA must have possessed these before it evolved into the two most ancient kingdoms of life: Bacteria and Archaea.

A cell membrane is fundamental to life as it is needed to contain and hold in all of a cell's chemical reactions, but they are very different within archaea and bacteria. Recent studies suggest LUCA had a leaky cell membrane (which modern life could not survive with), which allowed small hydrogen ions to pass through it while keeping the cell contents inside. It could have lived in ancient seawater where liquid dense with protons or hydrogen ions mixed with warm alkaline fluid from hydrothermal vents, which had fewer protons. The difference in concentrations of protons between the seawater and hydrothermal fluids allowed these hydrogen ions to flow into the cell, which led to the production of *ATP*. This energetic molecule transfers energy through a cell, powering its growth. Life would be now technically alive. The LUCA was, however, still stranded at the bottom of the ocean. To spread to new localities and even risk a journey to the surface it would need to evolve less leaky, stronger membranes to survive the less favourable environments. This is when bacteria and archaea would have started going their separate ways, each tweaking its membranes to make them less leaky, allowing them to set sail and colonise the Earth. And, of course, eventually to combine and evolve to become us!

When?

Does the first evidence of life date to 3.85 billion years or 3.45 billion, or even earlier? A 400-million-year discrepancy may seem trivial when discussing an event that happened almost 4 billion years ago yet scientists continue to argue about whether some of the oldest life-hosting rocks ever found date to 4.2, 3.85, 3.65 or only 3.45 billion years ago. The discrepancy matters because the rocks, however old they are, indicate that life already existed at the time they formed. So yes, as with everything else surrounding the origins of life – no one knows exactly when life began. Everything we *do* know is based on educated theories, carbon chemistry and some reasonably convincing wiggly-looking microfossils.

As mentioned before, very few rocks exist from the early Earth period and even if they did, prebiotic chemistry most likely would not have been recorded in them. As such, the oldest evidence that hints at life we have, that most scientists agree upon at least, is chemical. The ratio between the isotopes of carbon-12 and carbon-13 within graphite contained in the Isua metamorphosed sediments of Greenland is interpreted to represent traces of early life that flourished in Earth's oceans at least 3.7 billion years ago. If correct, this suggests that microorganisms were already well established on Earth by 3.7 billion years ago and so life must have started even earlier.

Microfossils found in Pilbara, Western Australia and Barberton, South Africa are suggestive of bacteria and even show enough variety to have been put into 11 different species. They are found in volcanic rocks dating back at least 3.45 billion years. Also, 3.5-billion-year-old dome-shaped rock structures, again found in South Africa and Australia, are suggestive of a microbial presence by this time. This is based on more than just fossils, since these stromatolites look remarkably similar to those that are forming today. Right now, communities of microbes, mainly photosynthesising cyanobacteria and heterotrophic (eats food for energy) bacteria that live in shallow, warm waters are producing stromatolites by forming thin microbial films that trap mud. Sheets of these mud/microbe mats build up into a layered sedi-mentary rock – a stromatolite. Scientists believe this is a similar process to how they would have formed billions of years ago, and to do it they needed some comparatively evolved microbial life.

Harnessing the Power of Sunlight

Although the exact timing of the origin of life is uncertain, what is clear is that by 3 billion years ago there was life on Earth, and plenty of it! Prokaryotes were widespread across the aquatic habitats of the Earth but were reliant upon and trapped within the environment in which they grew up, such as a single hydrothermal spring or black smoker in the depths of the ocean. The advent of the ability to utilise light from the Sun as an energy source to drive the synthesis of carbohydrates (sugars) from carbon dioxide and water – the process of photosynthesis – liberated life from these dark hidden recesses of the world and allowed it to visit the rest of the planet.

There is fossil evidence of the first photosynthetic bacteria 3.5 billion years ago, but because the Earth's atmosphere contained almost no oxygen during this time, many scientists believe that they did not generate oxygen as a waste gas in the way that photosynthesis does today, nor did they use visible light from the young Sun, instead using ultraviolet (UV) light. Early photosynthetic systems, such as those from green and purple bacteria, employed various molecules such as hydrogen sulphide as electron donors to obtain energy. This, however, required a steady supply of electrons from the surrounding environment, which limited their ability to move too far from the electron source. To become a fully mobile cell, life had to cut this umbilical cord and find a more renewable energy source. Life chose light. This allowed cells to be able to rebuild themselves using only carbon dioxide, water and sunlight: simple building blocks that are likely to be found on other terrestrial planets. As such, photosynthesis is quite possibly a universal process.

The Great Oxygen Predicament

On Earth, all atmospheric oxygen is produced through oxygenic photosynthesis; however, the only organisms capable of splitting water to do this are cyanobacteria. These first took hold in isolated marine and freshwater basins, producing only local oxygen oases, but hundreds of millions of years later, some 2.5 billion years ago, the first evidence of rising oxygen levels in the atmosphere and oceans was seen. This is called *the Great Oxidation Event (GOE)*. Iron minerals dissolved in the oceans acted as a sponge, initially soaking up the first free oxygen released by cyanobacteria. They formed red iron oxides that settled to the floor and over time hardened into sedimentary rocks that we now call banded iron formations. The surface of the Earth both above and below the waves was rusting. Once the Earth's crust had soaked up as much oxygen as it possibly could, the gas had nowhere left to go but up and the level of oxygen in the atmosphere rose rapidly. Apart from paving the way for oxygen-breathing life on Earth, the rising atmospheric oxygen levels had another important effect on life, especially that destined to live upon the surface. UV light from the Sun split apart oxygen molecules high in the atmosphere producing the molecule ozone, thereby creating the ozone layer. Ozone is a strong absorber of UV, so as long as there is oxygen in our atmosphere, sunlight will react with it to create a powerful shield against its own harmful radiation.

The Great Oxidation Event changed Earth's surface environment and made possible the evolution of large and complex life forms, including us. Yet, this event was also a *Great Oxygen Crisis*. The free oxygen produced can be a highly poisonous and toxic gas, especially for those anaerobic organisms that until this point were the dominant forms of life on the planet. The rising concentrations are actually believed to have wiped out most of Earth's inhabitants. Cyanobacteria were therefore responsible for one of the most significant extinction events in Earth's history. If that weren't problem enough, the free oxygen was also busy reacting with the greenhouse gas methane in the atmosphere, greatly reducing its concentration. Without this insulating greenhouse gas, the surface temperature of the Earth began to drop. The world was about to get really ... really ... cold.

Snowball Earth

Primitive humans clad in animal skins trekking across vast expanses of ice in a desperate search to find food – that's the image that comes to mind when most of us think about an ice age. But there have been several ice ages, most of them long before humans made their first appearance on the Earth. Our planet seems to have three main thermostat settings: *greenhouse*, when tropical temperatures extend to the poles and there are no ice sheets at all; *icehouse*, when there is some permanent ice, although its extent varies greatly; and *snowball*, when the planet's entire surface is frozen over. Between 2.2 and 2.3 billion years ago the Earth plunged into its first ever snowball event, during which the average surface temperature dropped and the ice advanced from the poles towards the Equator. Known as the Huronian glaciation, it is the oldest ice age we know about and is thought to have come about with the rise of oxygen and the negative effect this had on the planet-warming greenhouse gases. We do not know how far the ice grew around the Earth, but some liquid water must have remained as the unicellular water-dwelling life inhabiting the Earth was not completely extinguished. Perhaps the Earth was more of a slush ball than a snowball, with a thin equatorial band of open (or seasonally open) water.

Skipping ahead, there have been a number of big-freeze events that have engulfed the planet, not least between 750 and 600 million years ago in a time known as the Pre-Cambrian. They varied in duration and extent, but while in the grip of a full-on snowball event, life could only survive in ice-free refuges, or where sunlight managed to penetrate through the ice to allow photosynthesis to continue. It was a vicious cycle. Once cooling started, the growing regions of white ice reflected more and more of the Sun's warmth away from the planet, which created even colder conditions. As the ice encroached towards the Equator the seas sealed over, and thick ice blocked sunlight from reaching the photosynthesising aquatic organisms. Deprived of its source of oxygen, the seas turned anoxic and the Earth began to suffocate.

Each of these ice ages did end, thanks to the enduring hard work of the Earth and its volcanoes that, throughout it all, kept pumping their gases into the atmosphere. But the CO_2 cycle described in Chapter 3 was in trouble. Not only were the rocks that once trapped carbon dioxide hidden beneath the ice but the rate of photosynthesis dropped as well, so this greenhouse gas began to accumulate in the atmosphere up to 350 times its current level, with no way for it to be removed. However, the powerful insulating effect this rise in gases created started to warm the planet and after several million years it began to thaw. The climate of the Earth slowly settled back into an equilibrium and life returned to its previous niches. It is thought in fact that the release from such an inhospitable period actually provided the evolutionary jump-start for the seemingly sudden appearance of multicellular life on the Earth. Apart from the Andean–Saharan ice age from 460 to 430 million years ago, and the Karoo between 360 and 260 million years ago, Earth has been relatively free of frozen episodes from the Pre-Cambrian to today. This is quite possibly an important factor in worldwide evolution and the spread of life.

An Explosion of Life

About 600 million years ago the Ediacaran fauna arose and have now been found on all continents except Antarctica. The earliest known complex multicellular organisms, these were strange tubular or frond-shaped organisms, most of them sessile (not able to move from their position). They flourished until the cusp of the Cambrian 542 million years ago, when the characteristic communities of fossils vanish from the geological record. This is almost certainly because around half a billion years ago the Earth witnessed an evolutionary flare when in a short period of time almost every major animal group that has ever existed came into being. Before 580 million years ago, most life on Earth was simple, but in less than 80 million years (a mere blink of a geological eye) the diversity of life came to resemble that of today. This dramatic biological bonanza of evolutionary changes is known as the *Cambrian explosion.** The lower boundary of this period is widely agreed to be at 543 million years with the first appearance in the fossil record of worms that made horizontal burrows, whilst the end is marked by evidence of a mass extinction event about 490 million years ago.

During the Cambrian Period, the hypothetical single supercontinent Rodinia broke apart and by the early to mid-Cambrian there were two: *Gondwana*, near the South Pole, was a supercontinent that was the ancestor of much of the land area of modern Africa, Australia, South America, Antarctica and parts of Asia; *Laurasia*, nearer the Equator, was composed of landmasses that currently make up much of North America and the northernmost part of Europe. There were no continents located at the poles. In the early Cambrian, Earth was generally cold but was gradually warming as the glaciers of the many Snowball Earth episodes receded. The global climate ultimately became warmer than it is today and there were essentially no polar or high-altitude glaciers. Oxygen levels were only some 10 per cent of what they are today but, significantly, there *was* oxygen. The environment was becoming more hospitable for complex life and sea levels were rising due to glacial melting, flooding low-lying coastal areas and creating shallow marine habitats ideal for spawning new life forms.

At this point, no life yet existed on land; all life was still aquatic. The sea floor was covered with oozing mats of microbial life living on top of a thick layer of oxygen-free mud. The first multicellular life forms evolved to graze on these microbes and were themselves only

* Cambria was the Roman name for Wales. Since this region was the original study location for sedimentary rock formed during this interval of Earth history, the name 'Cambrian' was adopted for this time period.

near-microscopic worms that burrowed into the ocean floor, mixing and oxygenating the mud. They were also the first organisms to show evidence of a bilateral body plan that we still use today. The transition of pre-Cambrian life (mainly soft-body impressions in rock) to Cambrian life (shell-bearing fossils and other fossils with hard parts) was revolutionary. Among the animals that evolved during this period were the hard-bodied brachiopods, which resemble clams and cockles; arthropods with jointed external skeletons, the ancestors of spiders, insects and crustaceans; and the chordates, to which vertebrates (animals with backbones) including human beings belong. These toughened-up creatures represented a crucial innovation: hard bodies offered animals both a defence against attack and a framework for supporting greater body sizes. As such, many weird and wonderful forms of life came into existence during this time, although unfortunately few have survived through to today.

To explore the menagerie of alien-like creatures birthed during the Cambrian, we turn to rocks, the record keepers of Earth's history, and although rare and hard to find, the oldest rocks and minerals can provide a wealth of information about our past. However, there is a problem: most animals from the Cambrian had no hard parts such as bones or teeth and their soft gloopy bodies rarely left a fossil trace within rocks – the majority of fossil mammals are known only from their teeth, since enamel is far more durable than flesh or bone. Soft parts of bodies can only be preserved by a stroke of luck and commonly in an unusual serendipitous geological situation, such as amber leaking from trees that traps the fleshy bodies of insects before they get the chance to decay and disintegrate. Luckily this has happened in Pre-Cambrian and Cambrian sediments found in key areas of the world – such as Canada, Greenland, China and the UK – and has yielded a fantastic treasure trove of soft-bodied creatures. In fact, the first ever recorded discovery of *Charnia masoni*, the earliest known large, complex fossilised species on record, lay within the rocks of Charnwood Forest in Leicestershire, UK, and remains the only place in Western Europe where these ancient fossils have been found. These rarely located *Lagerstätten* or storage sites of squidgy life are possible thanks to a quick burial within sediments that were free of oxygen, which halted decay; provided protection from oxygen-breathing scavengers that would have consumed them; and kept them safe from the later ravages of the Earth, such as heat, erosion, tectonics and pressure.

The most famous fossil mother lode is found in Yoho National Park, British Columbia, Canada – a place more commonly known as the Burgess Shale. Found at around 2,500m (8,000ft) high on a mountain face above Emerald Lake in the Canadian Rockies this site, called Walcott Quarry, occupies one of the most majestic fieldwork spots I have ever been lucky enough to visit. In a lens-shaped bed of shale (the Phyllopod Bed) no more than 3m (10ft) thick and 60m (200ft) long, we have learned more about life during the Cambrian than from anywhere else in the world. The animals here probably lived on mudbanks built up at the base of a massive reef of calcareous algae (the reef-building corals we see today had not yet evolved). Mudslides could have dragged these ecosystems down into nearby basins that were deprived of oxygen, killing these organisms instantly. How do we know they died quickly? First, in the presence of an anaerobic, oxygen-free environment, marine invertebrates normally curl up upon dying. Fossils of the Burgess Shale do not exhibit this coiling; there was not enough time and so their death must have been quick. Second, there is no evidence of any attempt by these organisms to try to burrow out of their mud prison. If they had survived the fall, then surely they would have tried to escape.

The fossils of the Burgess Shale are spectacular, and many of them preserve exoskeletons, limbs and even guts. In some rare examples, there is actually 500-million-year-old evidence of stomach contents and muscle. In these rocks, the earliest known chordate (spinal cord-bearing animal) *Pikaia* was first found. Other marine creatures of Cambrian seas included the archaeocyathids and stromatoporoids (two extinct sponge-like organisms that formed reefs); primitive sponges and corals; simple pelecypods (ancestors of modern bivalves such as clams, oysters and mussels) and brachiopods; other simple molluscs; primitive echinoderms and jawless fish; nautiloids; and a diverse group of early arthropods. The iconic arthropods of the Cambrian were the trilobites, of which there is a huge number of fossils (there is one on my desk, in fact). Trilobites had flattened, segmented and plated bodies that helped protect them in seas that were increasingly filled with predators. With many varieties and sizes – they ranged from a millimetre to more than 50cm (20in) in length – and proved to be among the most successful and enduring of all prehistoric animals. Some species of trilobite were the first organisms to develop complex eye structures. More than 17,000 species are known to have survived until 251 million years ago.

Many species we observe from this time could have been stolen straight from science fiction. An example is *Opabinia*, a slim segmented animal with gills, five stalked eyes, and a long, flexible, hose-like structure extending out from under its head, ending in a claw fringed with spines. Another is the infamous *Anomalocaris*, which resembled the rear end of a shrimp. This gigantic predator was segmented with two large grasping appendages, a mouth with rings of razor-sharp teeth and was up to 2m (6.6ft) long. A personal favourite is *Hallucigenia*, so named because it looked so bizarre. It was a worm-like animal that walked on a set of 14 long rigid spines and had a row of tentacles along its back … or … did it walk on the tentacles and use the spines along its back for armour? This creature is so alien to us today that we cannot even determine which way up it goes.

Out of the Sea and on to the Land

One of the most important evolutionary steps for life was the greening of the Earth, but adapting to life on its rocky surface was an incredible challenge. Organisms needed to avoid drying out, having always been wet, and anything above microscopic size needed to create special structures to withstand the effects of gravity (which was previously stemmed by the buoyancy of the seas). Their respiration and gas-exchange systems also needed to change and even reproduction could no longer depend on water to carry eggs and sperm towards each other ... sex had to be re-invented.

When plants and animals began to transfer from the water to the land, the first organisms to lead the way were algal mats that dotted themselves along the edges of seas and lakes. This is because until this point, soil – a blend of mineral particles and decomposed organic matter – did not exist either. Land surfaces would have been either bare rock or unstable sand produced by weathering of the rock, and very dry. Microbial mats of photosynthesising cyanobacteria may have been the only organisms capable of survival, since today they are found in areas of modern deserts that are home to little else. True land plants are thought to have evolved from a group of branched, filamentous green algae dwelling in shallow fresh water, perhaps at the edge of seasonally desiccating pools, more than 470 million years ago. Soil-dwelling fungi were probably involved and formed mutually beneficial, symbiotic relationships with early land plants to assist them in their initial colonisation of terrestrial environments. Spores of land vegetation from non-vascular plants that lacked deep roots, just like mosses and liverworts today, have been found in Middle Ordovician rocks dated to some 476 million years ago. The terrestrial world offered these primitive plants mineral resources and plenty more exposure to sunlight than could be found in the crowded seas.

To survive on the land, plants had to become internally more complex and specialised. They needed to photosynthesise to provide food for the entire plant body and this was most efficiently conducted from the top; roots were used to extract water from the ground and the parts in between became support and transport systems for water and nutrients. The Middle Silurian rocks of around 430 million years of age contain fossils of actual plants, including mosses, but most were less than 10cm (4in) high. By the Late Devonian around 370 million years ago, ferns and trees such as *Archaeopteris* were abundant. The establishment of a photosynthesising land-based flora caused oxygen levels in the atmosphere to rise even further, and once it got above 13 per cent there was enough oxygen around to stoke wildfires. This is first recorded as charcoalified plant fossils. Apart from a controversial gap in the Late Devonian, charcoal has been found throughout the geological record ever since.

As the once barren continents became lush green land masses, a hospitable environment – and tasty food source – was finally available to support the first terrestrial animals. Various types of arthropod, the ancestors of millipedes and centipedes, the earliest arachnids, and the ancestors of insects came first. These ate the early plants and each other. Arthropods were pre-adapted to colonise land, because their jointed exoskeletons provided protection against drying out, support against gravity and a means of locomotion that was not dependent on water. Animals had to change both their feeding and excretory systems for life on the surface, and most land animals developed internal fertilisation of their eggs. If that wasn't enough the difference in refractive index between water and air also required big changes in their eyes. In some ways though movement and breathing became easier, and the better transmission of high-frequency sounds in air encouraged the development of hearing. The oldest known air-breathing animal is *Pneumodesmus*, an archipolypodan millipede from about 428 million years ago, but in general the fossil record of major invertebrate groups on land is poor. It is thought that insects developed the ability to fly in the Early Carboniferous, giving them a wider range of ecological niches for feeding and breeding, and a means of escape from predators. Then finally, we arrive at the *tetrapods* – vertebrates with four limbs – who evolved from lobe-finned fish over a relatively short timespan during the Late Devonian, 370–360 million years ago. As the continents continued to rearrange themselves into the continental land masses we recognise today, plants grew taller and evolved wooden stems, flowers and fruit. At the same time, vertebrates diversified from fish to amphibians, reptiles such as dinosaurs, birds and mammals, and finally nature made way for conscious intelligence.

The Stressors of Life

A sad fact of life on Earth is that without extinction events there would not be any life as we know it. Death is a necessity of life. We derive the history of life on Earth from the study of fossils and the rocks that contain them – significant events marked by an organism's first appearance and also its last. As mentioned above, multicellular life might only have been made possible by the release of the Earth from a freezing slush-ball period. In the last million years, throughout the Quaternary period, the Earth has undergone cycles of ice ages, each lasting about 100,000 years, yet the temperature difference has been less than 10°C (50°F). Earth is only now emerging from the last ice age that ended about 11,000 years ago and this coincidentally marks the cultural development of humans, which started only 10,000 years ago. The Earth's climate obviously had a huge impact on life and today we are still very much aware of its power over us. The fossil record tells us that since the Cambrian explosion there have been five major and almost catastrophic extinctions that have stressed life on Earth, but this time they had nothing to do with ice …

The major extinctions of the last 500 million years bear witness to the repetitive reboot of Earth's biosphere. The greatest is the Permian catastrophe 252 million years ago, whereby within one million years 70 per cent of all land species and 85 per cent of all marine species were erased from existence, but the jury is still out on what caused it, with both terrestrial and extraterrestrial culprits proposed. The most iconic mass extinction occurred 65 million years ago and is commonly accompanied by the image of a luckless tyrannosaur looking over its shoulder at a colossal fireball sent from the heavens as it streaks across the horizon, the monster's death by vaporisation imminent. The disappearance of most of our beloved dinosaurs, and actually 70 per cent of other species as well, although regretful, paved the way for the age of mammals and the eventual appearance of humanoids. This extinction is also popular as it has been linked to a powerful space rock 15km (9 miles) across slamming into the Earth. Sometimes comets and asteroids are forced from their orbits and head on a collision path with a planetary body. When these strike a world full of life, the dangers can be hard to comprehend and the devastation absolute. Even though there is, compared to the Moon, scant evidence of past strikes on the Earth, this is owing to the Earth's effective cleaning protocols of weathering and erosion erasing the evidence rather than the planet somehow avoiding such catastrophes.

A buried impact crater 180km (110 miles) wide of just the right age has been found in the Yucatan Peninsula of

Mexico, dubbed Chicxulub. The dinosaurs would not have been killed by this impact itself but rather by the environmental devastation that followed. This event would have triggered tsunamis across the oceans, caused powerful earthquakes and released enough heat to start spontaneous fires around the world. Material thrown into the air would have fallen back to Earth as acid rain thereby acidifying the oceans, and the dust would have blocked out the Sun, plunging the planet into a cold darkness for many years. Around this time, in an unrelated event, huge areas of western India, now called the Deccan Traps, were being smothered in lava, in some places more than 1.6km (1 mile) deep, while huge quantities of greenhouse gases were being pumped into the atmosphere. Asteroid-theory fans have long dismissed this volcanism as an irritating coincidence but many scientists now lean towards the idea of a planet weakened by overzealous volcanoes and then crippled by an asteroid – or vice versa.

There is a number of events we know of that could have dramatically altered the path of human life on Earth. In June 1908, in Tunguska, Siberia, 80 million trees were found burnt or flattened over 2,150km^2 (830 square miles) with no immediately apparent cause. It is now thought to have been the result of a small comet or meteor that vaporised mid-flight about 5–10km (3–6 miles) from the ground. A short time after the explosion, its noise and resulting air pressure fluctuations were recorded as far away as London, while dust rose into the stratosphere and reduced its transparency for months. Luckily, this impact occurred in a sparsely inhabited location; if it had hit near a city the results would have been ruinous. A bolide – a meteor that explodes in the atmosphere – the size of Tunguska is estimated to strike the Earth on average only once every 300 years, but we are always aware of the looming presence of nearby asteroids and were reminded of our vulnerability most recently in 2013.

On 15th February 2013, a meteor classed as a superbolide fell to Earth over Russia, temporarily outshining the Sun. Eyewitnesses felt its intense heat as it burned through the atmosphere at a speed of up to 69,000km/h (42,900mph). Thankfully, owing to its high velocity, shallow angle of entry, and our wonderfully thick protective atmosphere, the object exploded in an airburst over the city of Chelyabinsk. The explosion released 20–30 times more energy than that from the atomic bomb detonated at Hiroshima, generating a bright flash, a large shock wave and a hot cloud of dust and gas. Many fragments pelted the surface – fortunately, there were no resulting fatalities. The explosion injured around 1,500 people, mainly from broken glass from shattered windows, and damaged some 7,200 buildings in six cities across the region. Completely unrelated, a predicted and even larger asteroid approached close to Earth that same day, the roughly 30m (100ft) 367943 Duende, which passed quietly by some 16 hours later. What is worrying is that the Chelyabinsk meteor arrived undetected before it made contact with our atmosphere. Measuring around 20m (65ft) in diameter, it is the largest known natural object to have entered Earth's atmosphere since the event at Tunguska.

We have managed to summarise more than 4 billion years of evolution in this chapter, but when we think in terms of astrobiology and the search for life on other worlds we are not expecting to find any life resembling that which has arisen here in the last 400 million years or more. It's the simpler life and the curious forms it took early in its history that can teach us the most. Thanks to our, albeit still limited, understanding of evolution on the Earth we may one day recognise alien creatures as similar to one of the many wondrous forms of terrestrial life that has existed though may not be present today. Maybe other planetary bodies have life but it has only evolved to the point of the Cambrian, or is only just starting to grow plants, or is in the midst of an age of dinosaurs. Our terrestrial ancestral aliens can teach us so much about the elaborate or painfully simple forms that life could take on another world, and in doing so keep our eyes and our minds open to the unexpected.

CHAPTER FIVE

Alien Worlds on Our Doorstep

The search for life in the Universe is not an easy pursuit. There is such a vast repository of information about the many wondrous forms life has taken over the history of the Earth that it is theoretically possible (perhaps almost inevitable) that life on another world could take on any number of guises as it would be affected by and respond to its particular environment. At the very least we have some educated ideas about what to look for and what types of environment on the Earth they like to inhabit. Now we just need to figure out where to look in the cosmos. Planets, moons, asteroids and comets are all options but they are not small targets. The search needs to be narrowed down to particular environments or geological features that hold the greatest chance of supporting life forms. The Earth, as with life, has an incredible selection of sites from which we can choose and learn from.

A Home Away From Home

Ultimately, the search for life in the Universe begins by looking for localities where all the conditions needed for life to exist can be found together, in the exact same spot. This is called the search for *habitability*. Habitability requires a set of physical and chemical conditions, such as the availability of water, energy and carbon, that if found would first give life the opportunity to become established; and second, sustain it and allow it to flourish over the longer term. Unsurprisingly, any location that displays such 'must-have' parameters is termed a habitable environment and it is these areas astrobiologists are most interested in finding. One such environment is the Earth itself: a perfect example of a globally habitable world, with billions of smaller habitable niches present within it.

The Earth is unique and incredibly special. Bringing together everything we have learnt about its make-up in previous chapters we can encapsulate our planets' suitability for life. More than 70 per cent of its surface is covered in liquid water – some might say the most important prerequisite for life. Unlike any other world in the Solar System, it is able to support this liquid water because of its thick insulating atmosphere with a greenhouse effect that prevents wild swings of temperature, and keeps the water from globally freezing or boiling away – the greenhouse effect is not always a negative thing. The atmosphere both generates an oxygen-rich ozone layer that absorbs harmful UV radiation from the Sun, and serves as a barrier to protect the fragile life below. The Earth also has a global magnetic field, something that other planets such as Mars have long since lost. The magnetic field protects the planet's atmospheric bodyguard from being blown away by solar winds, and additionally provides another line of defence against cell-damaging cosmic radiation. Finally, the Earth is a rocky or terrestrial planet built mostly of silica-rich rock instead of metal or gas. This allows plate tectonics to be sustained, with this process crucial for the long-term sustainability of Earth's climate and life. In the search for habitable environments on a global scale, astrobiologists are trying to locate a world similar to the Earth – looking to rocky terrestrial planets and moons with some element of an atmosphere and magnetic field.

The search for life and habitable environments so far has focused on finding a world with water, a source of energy (to power life) and a source of organic carbon. This is, of course, because all terrestrial organisms observed so far are reliant on liquid water for their survival, although fluids other than water have been found on other planets and these may have the potential to support life (see Chapter 8). Life and therefore life-friendly environments also need energy to drive and sustain metabolic processes and encourage growth and reproduction. Ultimately, this energy comes from the Sun, even though it is 149,600,000km (almost 93 million miles) away, and is readily available to any organism or habitat located on or close to the surface of a planet or moon. If conditions do not allow for life on the surface, however, then habitats may be found lurking underground where chemical energy can be used as a replacement for solar. This process sees microbes break complex compounds into simpler ones to obtain just a small amount of energy from the chemical reaction. Finally, a habitable environment needs organic carbon-based molecules – the building blocks for life. These are not necessarily biological (as they can exist without being part of or created by an organism), but life cannot exist without them. Luckily for life, they are found in every corner of the known Universe, within meteorites that have pelted planets and moons for billions of years, and in comets travelling through our Solar System and throughout the interstellar medium. This is extremely important, as the universal nature of carbon molecules and their delivery across the cosmos greatly increases the habitability potential of millions of worlds across our Galaxy and beyond.

This appears a fairly simple formula – locate carbon, water and energy on a terrestrial world or even within a single palm-sized rock – and you will find a habitable environment flourishing with living organisms. However, nothing is ever that simple. First, think about the relative nature of the term *habitable* and how it changes depending upon the type of life you are looking for. A boiling acidic hot spring in Yellowstone National Park might be the perfect home for a heat-loving bacterium yet it would be fatal for a human. In fact, no world yet discovered in the Universe would be entirely suitable or even remotely habitable for a human being without extensive artificial help. Most have such extreme environmental conditions that it is almost inconceivable that any form of terrestrial-like biology could exist there. Second, environments are not static as they fluctuate and change over time. The conditions present during the origin and first appearance of life are not necessarily those needed to maintain the life forms that are created, or even capable of supporting their long-term survival. Mars is a good example of this. Mars it's believed may have had a warmer, wetter, life-friendly environment, but today is an inhospitable frozen wasteland. Life that might have arisen early in Mars' history may not have been able to survive as its environment degraded and froze. Third, science is making some mighty assumptions. It assumes that any location that has, or has had in the past, the ability to support life will definitely have contained it. It has to be considered, however, that there are environments where there is no life in spite of there being ideal conditions for it – so-called uninhabited habitats – and can range in size from a single blade of grass to an entire planet. At present, astrobiologists are searching the Universe for environments that have the potential to support life either now or in the past, but may not prove to be inhabited. We are not just searching for conditions that we as humans would thrive in, of course, since this would be a very short and ultimately futile venture. Rather, we are hunting for environments that push the limits of biological survival to its very extremes, acutely aware that life has the resilience to become established and to survive in some incredibly unusual places.

Alien Environments All Over the Place

When searching for habitable environments, it goes without saying that we have to think about the type of life that might be able to live in them. So in locating potential habitats in the Solar System, it is important to consider the physical limits of life and the constraints this places on a suitable home. The main defining factors affecting cell-based life are temperature, acidity and salinity. Excitingly, certain terrestrial life would be perfectly content in extremes of these conditions and such environments are commonplace on extraterrestrial worlds. Wonderfully, these extreme locales can be found in hundreds of pockets across the Earth that we can visit.

Figure 3 *Terrestrial analogue sites. From left to right: Rio Tinto river, Spain; Geysers, New Zealand; Salt playas, Spain; Eyjafjallajökull, Iceland. (credit: Louisa Preston).*

Alien worlds and even alien life forms can essentially be studied right here on our doorstep in places called analogue environments, as their biology, geology, chemistry or physical appearance (or a combination of all four) mimic an environment that was once found or currently exists on another planetary body. At present, terrestrial analogue studies are the best way for scientists to examine the habitability potential of alien environments, and they are able to help us design and develop tools and technologies for their exploration. No analogue environment is ever a perfect replica of a location or the conditions present on another world, however. For example, there is nowhere on Earth that naturally mimics the different forces of gravity found on other planets and moons, nor usually their atmospheres. The analogues we have do, however, display a number of extraterrestrial features that can be compared with those on other planetary bodies, to try to understand them better.

Some habitable analogue environments are considered more significant for exploration than others. These are based upon direct observations of their existence on other worlds through space missions, by orbiting satellites and from Earth-based telescopes. For example, we have data and images to certify that Mars has volcanoes, is composed of volcanic rocks similar to those found in Iceland or Hawaii, is covered in impact craters and has ice caps and glaciers, which are all possible sites where life might be hiding, as it is on Earth. This means that volcanoes, basalts, impact sites and ice-covered habitats on Earth are all extremely important analogues, albeit never perfect ones.

Other analogues are based on theories and circumstantial evidence. These are identified by indirect or highly suggestive evidence of their existence, which is still awaiting data to provide confirmation of the theory. Research into the conditions and biota within Lake Vostok in Antarctica is a perfect example, as it is used by scientists to explore the habitability potential of a brine ocean that may lie under the icy crust of the Galilean moon Europa.* There is abundant indirect evidence to support the existence of this ocean and its possible composition, but we have yet to prove that it is there. Another example is the lack of flowing rivers of liquid water across the surface of Mars. Evidence of past water action and seasonally stable liquid water is globally present in the form of features that look identical to dried-up river channels, deltas, flood plains, lakes and seas on the Earth. Such indicators would include dark recurring slope lineae (RSLs) and specific minerals that can only form in the presence of water. While the connection between a river channel and liquid water seems almost undeniable, direct evidence for permanently flowing liquid water is still hard to find.

Finally, we have the analogues that are not backed up by any scientific evidence whatsoever. Unconfirmed UFO sightings aside, there is no physical evidence that life exists anywhere other than on Earth; nonetheless, terrestrial analogues for a mythical alien life form are incredibly important and are integral to the development of scenarios for planetary habitability. The premise that because life is found on Earth in a specific niche environment, and that a similar environment present on another planet could therefore be a habitable environment for life, is perhaps a reasonable one to maintain. The existence on Earth of extremophiles, extreme environment-loving organisms described in the next chapter, is an example of this life. A combination of this analogue research with laboratory experiments and data collected from the Earth itself and other planets and moons is allowing astrobiologists to develop some reasonably educated guesses as to where to look for life.

---

* One of three Jupiter satellites discovered by Galileo.

Extreme Living

The more we explore other worlds, the more we see that their environments are extreme in comparison to those on the Earth. As such, when looking for places on our planet where we can study life forms that might be able to thrive elsewhere, we turn to those localities that depart from our norms, that house conditions where we as humans would struggle to survive without help. These include hot springs and geysers, deep-sea hydrothermal vents, hypersaline environments such as salt flats, deserts both hot and cold, glacial ice, evaporites (sediments precipitated when water evaporates), and even the atmosphere itself. Instead of describing each type of environment, it is more meaningful to show how certain places on the Earth function as analogue sites for some of the most astrobiologically attractive worlds in the Solar System.

A Home Fit For a Martian

In the search for alien life, one instantly thinks about the possibility of life on Mars. This single Solar System body is the most Earth-like planet we have seen and as a bonus we have fairly good access to it. I believe we will one day find life there, or at the very least evidence of it having once existed. Owing to their similarities, the Earth boasts hundreds of analogue environments that mimic closely not just the Mars of today but also the various environmental conditions it has experienced over its 4.5-billion-year existence: as it changed from a warm and wet oasis to a cold and arid desert. One of the main steps in assessing the habitability of Mars, therefore, is to study a similar range of Earth-based environments, understand what makes them habitable and what life forms can exist in them, and unearth their level of dependency on water – this is what I do! Needless to say, establishing the presence of life on Mars is a huge challenge, because as the availability of liquid water on the surface of Mars has fallen, so too has the planet's habitability potential and sadly the chances of it hosting life today.

The early years of Mars could be considered an Eden for life, as liquid water was abundantly available on its surface, temperatures would have been higher and its thicker atmosphere would have protected the planet. Rio Tinto, located in southwest Spain, is a fascinating analogue for habitable environments and the life forms that may have existed on Mars during this period of its history. Rio Tinto is a natural acid-rock drainage system, flowing with blood-red, iron-rich waters that teem with life. The pH of the river is an extremely acidic 2.3 (the same acidity as if it were flowing with lemon juice), directly created by its microbial community. Organisms here are chemoautotrophs that derive energy from chemical reactions between inorganic compounds such as iron (a process called chemolithotrophy), with iron-oxidising, acid-tolerant filamentous bacteria such as *Leptospirillum ferrooxidans* and *Acidithiobacillus ferrooxidans* dominating it. Rusty iron-rich terraces of rock have been forming along the banks of the river for more than 2 million years, and these have trapped bacteria and other microscopic organisms inside them. This site is a wonderful natural laboratory where living organisms can be observed and then matched to their fossil ancestors in the surrounding rocks. This naturally gives rise to the questions 'Might these or similar types of organism once have lived in ancient rivers on Mars?' and 'Could we find evidence of organisms like these preserved in rocks on Mars?'

Time has taken its toll, however, and the Mars of today has lost most of its surface liquid water, has begun to rust and has become a dusty frozen desert. The most Mars-like environment on the Earth today is found at the South Pole, in the vast white Antarctic desert and the spectacular Antarctic Dry Valleys. These are the coldest and driest regions on the Earth, and mimic the harsh arid conditions now prevalent on Mars. Temperatures in the Valleys plummet to −40°C (−40°F), which, combined with the highest UV-B radiation levels on the planet and no source of liquid water, create extreme environmental stresses for life. On first inspection they appear completely barren. Yet within glacial ice and sub-surface rocks, life flourishes. Buried under the surface are warmer wet niches that create microscopic habitats for algae, fungi, nematode worms and tardigrades (the water bears whom we will meet in the next chapter). These endolithic (within rock) realms give hope that similar hidden habitable environments could persist on Mars today and may have provided a refuge for life when conditions on its surface started to deteriorate.

Observations and investigations on Mars have driven the need for greater analogue research which has in turn allowed for the discovery of more habitable environments, not just on Mars but also throughout the Universe. This has meant the discovery of hundreds of extreme sites on Earth that have drawn attention to new, potentially habitable patches of other planetary bodies that had been previously overlooked and are even providing an insight into the origins of life itself.

Floating above Venus

On the opposite side of Earth is an unlikely target in the search for habitable environments in the Solar System: its sister planet, Venus. It is common knowledge that Venus is pretty inhospitable for life, now known to resemble our clichéd imaginings of hell rather than a lush tropical paradise. It has the highest surface temperature (an unimaginable 460°C/860°F) of any planet in the Solar System (including that of Mercury), therefore liquid water is a near-impossibility on its surface. It also has a toxic, unbreathable, carbon-dioxide-rich atmosphere and clouds dripping with sulphuric acid. The Venusian surface is not a habitable environment for life as we know it. High above the sweltering ground, however, is a potential habitable zone where temperatures lie between 0°C (32°F) and 120°C (248°F), and water vapour is available. The lower and middle cloud decks within this high-altitude region may therefore support an aerial biosphere. These clouds offer long-lasting droplets of water, although they are highly acidic and rich with dissolved hydrogen sulphide. Energy sources are available through chemical reactions such as sulphate reduction, or indeed photosynthesis, as happens in plants on Earth. Acid-loving organisms found in hot acidic waters on Earth, such as the bacterium *Acidianus infernus*, which grows at 88°C (190°F) and in acidic conditions down to pH0.5, are key analogue organisms for Venus. In addition, there is a precedent set for floating life, from Earth. Evidence shows that bacteria may be actively metabolising, and even reproducing, in clouds way above the Earth. So why not on Venus?

Although most similar to Earth in build and distance from the Sun, the terrestrial planets of the Solar System do not present much in the way of enticing habitable environments on their surfaces today. As we've seen, our strongest hopes lie with Mars that it may contain preserved biosignatures of past life forms within surface rocks and minerals, or active communities enduring deep in the sub-surface. As such, the traditional formulation of the habitable or Goldilocks zone around a star is perhaps too restrictive as we start to understand just how far life can and might need to go to survive. A planet or moon may not need to be a similar distance the Earth is to the Sun from its host star to house environments suitable for life. Habitable environments existing throughout the Solar System and in other solar systems across the Universe are becoming a possibility.

The Jovian Moons

Surprisingly, some of the likeliest candidates for life-hosting habitable environments in the Solar System are to be found on moons. These are becoming more important in the search for life than the planets they orbit. The most promising are found among more than 210 frozen natural satellites orbiting the gas giant worlds of Jupiter and Saturn. Recent space missions have revealed many of these moons to be geologically active bodies, with volcanoes spewing ice as well as molten lava, geysers the size of whole countries on Earth, impact craters in their thousands, and vast channels and valley networks. Excitingly, these moons are displaying a wealth of potentially habitable environments. The problem for astrobiologists, however, is that although these geological features have the potential to house alien life, the moons are so incredibly extreme compared to the Earth that suitable terrestrial analogues of their environments and potential life forms are harder to find. In addition, our knowledge of the conditions actually present on these icy moons is mainly based on inferences rather than on deductions supported by definitive data. There is a great deal more educated guesswork and imagination needed for detecting life on these worlds than those closer to home ... but that is half the fun.

The Oceans of Europa

One of the most important moons in the search for habitable environments in the Solar System is Europa. This moon is actually built like the Earth and the other terrestrial planets, mainly of silicate rock, but instead of a surficial liquid it is covered in a smooth sheet of water ice that is many kilometres thick. This icy shell is, however, believed to hide a secret ocean of briny water beneath. You might think that this ocean would be fairly small, but in fact Europa is only slightly smaller than the Moon, and the volume of its ocean is estimated to be 3×10^{18} cubic metres: this is twice the volume of all Earth's oceans put together. Europa's frozen temperatures ($-220°C/-364°F$ at the poles), resulting from its huge distance from the Sun, provide an extreme environmental challenge for life, although a number of ice-dominated habitats on the Earth could supply analogues for liveable environments here. Most significantly, there is an important analogue site for the ocean itself. Lake Tirez in Spain contains very salty, sulphate-rich waters that may be similar to Europa's salty interior. In addition, salt-loving organisms growing in these Spanish waters provide an insight into how a habitable environment could exist on or beneath Europa's surface.

The Earth also has analogues for the capability of life to survive buried under a shield of ice. Liquid water lakes hidden up to 3.2km (2 miles) beneath the ice sheets of Antarctica, such as Lakes Vostok, Ellsworth, Bonney and Vida, are thought to be similar to Europa's salty sub-surface ocean. Core samples taken from the ice surrounding Lake Vostok in 2012 revealed DNA from an estimated 3,507 organisms. Similar under-ice realms on Europa are believed to have the best potential to host microbial ecosystems in the entire Solar System, bar the Earth. A habitable sea-floor environment may also occur on Europa. There are extensive communities in the dark, cold, high-pressure environment of the Earth's ocean floor, particularly around deep-sea hydrothermal vent fields such as Lost City on the Mid-Atlantic Ridge and the Mariana Trench in the Pacific Ocean. It is important to investigate these analogues, even if at present any actual search for a deep-ocean biosphere on Europa is impossible. We instead are hunting for indicators of their activity and the presence of this ocean on the surface ice.

The Fountains of Enceladus

One of Saturn's many moons, Enceladus, has attracted great interest thanks to dramatic images of powerful icy jets currently erupting from more than 100 cryovolcanoes near its southern polar region. These geysers explode over 645km (400 miles) into space (about the distance from London to Paris!). Measurements taken of these gigantic plumes by the Cassini Orbiter found water gas, simple organic carbon-based molecules and volatiles such as nitrogen and methane to be present. These life-essential compounds must have come from source regions far inside the moon, from the area that feeds the jets. The assumption, therefore, is that organic molecules used by and needed for life are present deep within Enceladus. The young volcanic landscapes of Iceland provide a good analogue for these plumes. Iceland is covered in geysers and hot springs, cracks in the Earth's surface where near-boiling water erupts in spectacular fountains, blanketing the ground with mineral- and nutrient-rich waters. Surrounding the hot springs of Iceland are heat- and acid-loving bacteria that form mats of microorganisms, creeping across the surface and thriving in the hot acidic waters. Such terrestrial features are miniscule versions of the gargantuan jets seen erupting from Enceladus but they can inform us about the processes involved in their formation and their ability to create and support a habitable environment.

The Lakes of Titan

One place that greatly resembles the Earth in appearance is another of Saturn's moons, Titan. It is the only moon in the Solar System known to possess a thick atmosphere, and a substantial one at that, and there is evidence that it has Earth-like lakes and seas, rivers with running fluids, sand dunes and weather systems. One lake, Kraken Mare, is three times larger than Lake Michigan-Huron. Titan's lakes are crucial targets in the search for habitable environments and life on this moon. However, surface temperatures of around $-179°C$ ($-290°F$) suggests that the liquid bodies on the surface could not be composed of water, but are more likely to be a mixture of methane and ethane – hydrocarbons. Life might be present here within a range of habitats, from the liquid hydrocarbon lakes on the surface to great depths into the sub-surface, creating a potential biosphere volume double that of the Earth. Owing to the very different chemistries of the liquids on Titan, we can only speculate about what life might be like there, and as such there are very few analogues for this world. The best known one is Pitch Lake, on the island of Trinidad – a natural liquid hydrocarbon or asphalt lake just like those found on Titan, albeit a far smaller version. A unique microbial community is found here, one that includes archaea and bacteria that are actually anaerobic – they are able to live without any oxygen at all. The natural asphalt-soil seeps of the Rancho La Brea Tar Pits in California and the Alaskan Oil Field petroleum reservoirs are also potential habitable analogue sites for Titan.

The Signatures of Life

Once we find these habitable environments on Mars, Europa or Titan, and can send either robotic landers to hunt them down or even, one day, humans to investigate them, what will we look for? It is highly unlikely that organisms will be caught scurrying across the surface of Mars or swimming in the lakes of Titan (although never say never). We will instead be searching for the evidence that life leaves behind. These signs of life or *biosignatures* will be recognisable as they will be composed to a variable extent of carbon.

A biosignature is classed as any substance, be it an element, isotope or molecule, that provides scientific proof of past or present life. The usefulness of a particular biosignature is determined not only by the probability of life creating it, but also by the improbability of non-biological processes producing it. Life processes may produce a range of biosignatures such as nucleic acids (the building blocks of DNA), lipids (fats), proteins and various morphological or visual features that are detectable in rocks and sediments (think dinosaur bones, trilobites, ammonites and any other kind of fossil). In addition, life interacts with its surroundings; for example, it can cause changes through chemical reactions with rocks and fluids, altering their chemistry or creating new materials. These processes will leave features in the geological record that indicate that life was once present.

Biosignatures are commonly used in geochemistry, geobiology and geomicrobiology to determine whether living organisms are or were once present within samples. Now they are applied to astrobiological exploration, founded upon the premise that biosignatures encountered in space will either mimic those found on Earth or be undeniably recognisable as originating from extraterrestrial life. An example of such biosignatures might be complex organic molecules, or structures whose formation is virtually impossible without the help of life. Some categories of *space* biosignatures include: cellular and extracellular morphologies (fossils, in other words); bio-organic molecular structures (e.g. lipids, proteins); chirality (a molecule's left- or right-handedness affected by interactions with living organisms); the presence of biogenic minerals (such as opal, only formed by life processes); atmospheric gases (e.g. methane and ozone that are largely produced via biological processes); and remotely detectable features on planetary surfaces, such as photosynthetic pigments.

Many of these signatures will not be detectable without the help of rocks, minerals or ice to preserve and protect them over millennia. Encasement within these media allows for the preservation of the remains of living organisms to be studied after their death, and their identification after geologically significant periods of time (billions of years). In general, this occurs through the process of fossilisation. The most common method being mineralisation, whereby hard parts of an organism are replaced by minerals such as calcite, silica, pyrite and phosphate, as well as a number of clays that were dissolved in water present in the sediment in which the organism died, or fell into shortly after death. An environment must satisfy a number of criteria as a suitable site for fossilisation, and some of the most valuable habitable environments we have found on other planets would be excellent sites for this process. As fossilisation and replacement of the original organism progresses, cell contents, cytoplasmic details and wall structures can be destroyed, rendering the identification of the original organisms difficult. In addition, after millions or even billions of years encased inside rocks, these fossils can be broken down, cracked, rearranged, completely destroyed by tectonics, weathering and erosion, or buried so deeply that we might never find them. Ultimately, this makes it incredibly hard to find and recognise a fossil of a once-living organism. Molecular biosignatures therefore can be extremely valuable as they are more easily preserved within rocks, they can survive for much longer periods of time, and they can tough out a number of physical processes that would easily destroy a fossil. They are the key to the search for evidence of past life on Mars, Europa or any solid planetary body, and as such identification of analogue biosignatures in the ancient rock record on Earth is crucial; it provides an opportunity to learn which geochemical signatures are unequivocally produced by life, and how they are preserved over geologic time.

It is important to avoid a false positive result in the search for life. It would be a disaster to cry wolf over such a globally important discovery. Fossil-like objects may resemble once-living life forms, but they must be proven to be biogenic before claims of life are made. To find definitive evidence of living organisms on another world and to prove an object's biogenicity, we therefore need to observe the co-occurrence of biological morphology, *i.e.* fossils and carbon chemistry. Biological information needs to be extracted from any candidate life forms to prove that they were once, or are currently, alive. DNA as well as proteins and fatty acids are considered key pieces of irrefutable evidence of life. The use of analogue environments on the Earth, such as Rio Tinto, is therefore important in this search. Here, filamentous fossils that *appear* to be bacterial in origin are studied using a range of analytical techniques to identify proteins and fatty acids preserved within them. Bearing such markers, the fossils can be confidently assigned to life and are trustworthy evidence of previous habitable conditions in the area. Studies on fossil localities in Earth's oldest rocks, such as those in the Greenstone Belts of South Africa and Australia, have shown that fossils and their associated biosignatures can be preserved and identified. Sites such as these provide ideal testing grounds for discovering the best techniques for the identification of markers of life on Earth and on other worlds.

Exploring the Extremes

Finding life anywhere on another world is just one use of analogue sites. Another is to figure out how humanity itself might one day leave the Earth and survive on another planet or moon. As such, a good analogue site is also a location where the exploration conditions of future astronauts can be simulated. Future explorers of the Moon or Mars will need to handle various conditions, such as reduced gravity, radiation, extreme temperatures and working in pressurised spacesuits. Preparing astronauts calls for training on sites that exhibit some or all of these conditions. The operations that can be simulated extend from living in isolation, cooking and gardening, doing fieldwork in a spacesuit and extravehicular activities (EVA) in reduced gravity to the construction of future habitats for humanity.

In order to help develop the key knowledge required to prepare for human exploration of Mars, the Mars Society initiated the *Mars Analog Research Station* (*MARS*) project. A global programme of Mars exploration operations research, this project includes two Mars base-like habitats located in deserts in the Canadian Arctic (the *Haughton Mars Project*, *HMP*) and Utah (*The Mars Desert Research Station*, *MDRS*). In these Mars-like desert environments, extensive long-duration field exploration operations are conducted in a similar style and under many similar constraints as would occur on the Red Planet. MDRS is a laboratory for learning how to live and work on another planet and is a prototype of a habitat that could house humans on Mars and serve as their main base of exploration. The station serves as a home-from-home to teams of six or seven crew members, including geologists, astrobiologists, engineers, mechanics, medics, human-factors researchers, artists and others, who live for weeks to months at a time in relative isolation, as they would on the surface of Mars.

NASA's *Hawai'i Space Exploration Analog and Simulation (Hi-SEAS)* mission is another analogue site used to prepare for human space flight to Mars. Located on the slopes of the Mauna Loa volcano on the island of Hawaii, this isolated dome is surrounded by a Mars-like terrain and houses a crew of *terranauts* who are researching what is required to keep a space-flight team happy and healthy during an extended or even permanent mission to Mars. Research into food preparation and growth, and crew dynamics, behaviour, roles and performance is carried out by the team, who also must live their daily lives, do chores, conduct EVAs in spacesuits and contend with 40-minute delayed communications, as would happen on Mars. A global mission support team of more than 40 volunteers, including myself, provides 24/7 technical assistance and a friendly ear. The first mission in 2013 lasted for 4 months and in 2015 the first year-long mission began.

The *NASA Extreme Environment Mission Operations project (NEEMO)* is an analogue mission that sends groups of astronauts, engineers and scientists to live in Aquarius, the world's only undersea research station. Operated by Florida International University, Aquarius is located 5.6km (3.5 miles) off Key Largo in the Florida Keys National Marine Sanctuary. It is deployed next to deep coral reefs 19m (62ft) below the surface. Underwater analogue sites allow for the training of *aquanauts* in neutral buoyancy conditions while operating in a natural but extremely hostile alien terrain. The aquanauts experience some of the same challenges beneath the waves that they would on a distant asteroid, planet or moon. During NEEMO missions, they simulate living on a spacecraft and test spacewalk techniques for future space missions. The underwater condition has the additional benefit of allowing NASA to *weight* the aquanauts to simulate different gravity environments and a technique known as saturation diving allows the aquanauts to live and work underwater for days or weeks at a time. Potential targets for such training are missions to the International Space Station (ISS), the Moon, Mars and asteroids, to test sampling, drilling and field explorations in one-sixth or one-third of Earth's gravity and to test anchoring systems in microgravity. A slightly different type of underwater analogue site is based at the *Pavilion Lake Research Project* (*PLRP*) in British Columbia, Canada. Since 2004, two-week missions have been conducted every summer to train astronauts how to search for evidence of life in an extreme environment with reduced-gravity conditions – however, astronauts only do EVAs underwater; they do not live there.

There are three permanent, all-year research stations on the Antarctic Plateau: *Concordia Station* (French–Italian), *Vostok Station* (Russian) and the *Amundsen–Scott South Pole Station* (US) at the Geographic South Pole, the southernmost place on the Earth. One of the coldest places on our planet, and the world's largest desert, temperatures here hardly rise above −25°C (−13°F) in the summer and the lowest natural temperature ever measured was recorded at Vostok Station: a frost-shattering −89.2°C (−128.56°F). These stations conduct a great deal of planetary and astronomical research but, perhaps most interestingly, they all allow the study of stressors associated with long-duration space missions, including extreme isolation and confinement. During the winter, crews are without the possibility of evacuation or deliveries for 9 months and live for prolonged periods, up to 6 whole months, in total darkness. Concordia station has been proposed as one of the highest-fidelity, real-life Earth-based analogues for long-duration deep-space missions.

The possibility of finding life somewhere other than the Earth seems to increase the more we understand our own planet, the conditions in which life has been found to survive and thrive, and the more we see data from the orbiters and landers that we are sending to other worlds. Although the planets and moons of our Solar System may prove to be habitable, however, it still does not mean that there is life on them. Earth remains the only example we have of an inhabited planet where life originated and evolved from a single-celled organism to the plethora of species we see today. Through working in areas across the Earth that exhibit similar traits to places on other planets and moons in the Solar System, we know that we need to target habitable environments around the numerous impact craters, ancient volcanoes and sub-surface environments of Mars, within the salty liquid oceans beneath the ice of Europa, in the source regions of the gigantic water jets erupting from Enceladus and within the Earth-like hydrocarbon lakes and seas of Titan. Each of these environments is considered to be *extreme*, and will exert immense stresses and pressure on any organism, including humans, trying to exist within them. Even though the conditions for life are tough … life is tougher.

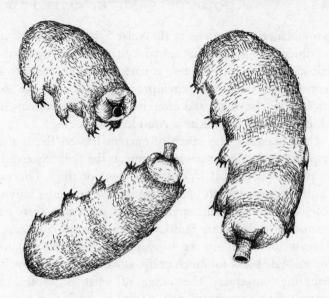

CHAPTER SIX

Everything is Relative!

One of the developments in recent years that really opened up scientists' eyes to the possibility of life on other worlds was the realisation of just how adaptable and versatile life is, and the growing appreciation and understanding of the physical limits of organisms. Since the 1990s, a particular branch of microbiology has been gathering pace – the area of *Extremophiles*. In every corner of our planet in which we look, even in places barely survivable by humans, there is some form of life flourishing. It seems that once life gets going, it will fight to survive and, if needed, will adapt to fill almost any moist niche it comes across. With evidence for liquid water surfacing on many planets and moons in the Solar System, although in conditions beyond any that humanity can survive, organisms that have adapted to life in extreme conditions are taking centre stage. They are providing us with a template for finding life elsewhere and directing us to environments in which we would never have considered searching.

Understanding the range of current life on Earth and mapping it to current environments in the Solar System is simply the start as it lacks the element of time. Life on Earth was substantially *alien* when it arose around 4 billion years ago because the environments on Earth were so dramatically different. Similarly, the climatic conditions forecast for a billion or so years into the future are depressingly bleak for much of life as we currently know it, including ourselves. The range of what is possible is continually being stretched to incorporate new adaptations displayed by life in its bid for survival. Some organisms have created an extreme-living club, and you can only be a member if you can survive somewhere another form of life cannot.

Extreme-lovers

Terrestrial life, as defined in Chapter 2, is made up of carbon compounds and uses water as a solvent and as such has to abide by the limitations imposed by this chemistry. It manages to emerge and survive within boundaries such as the boiling and freezing points of water, the presence or absence of oxygen, extremes of acidity and alkalinity and all pHs in between. The range of tolerances to these are known for the most part for all cells, and the extent to which life can survive any combination of acidity, temperature and salinity determines the *envelope of life*, ranging from cold acidic waters in one corner to hot alkaline brines in the other. The complexity of eukaryotic cells means that they are much more sensitive to perturbations of these three conditions. Therefore, the most extreme outer regions of the envelope of life are not dominated by complex eukaryotes but by the simpler prokaryotes. Any organism found thriving in these hostile and extreme environmental conditions on Earth is bestowed with the alias extremophile, literally meaning 'extreme love'. The extremophilic taxonomic range spans all three domains of life, including multicellular sophisticated vertebrates. While adaptation to a single harsh habitat is already impressive, there are species that can survive a variety: the polyextremophiles. These organisms are exploiting an ecological niche for which they are uniquely adapted, and face little or no competition within it.

The Relativity of Extremity

What is *extreme*? Perhaps extreme is simply an opinion; whether an environment or an organism is extreme is determined by the eye of the beholder. It is clear to us humans that a heat-loving thermophile that dies at temperatures below 21°C (69.8°F), or a pressure-loving piezophile that finds the atmospheric pressure on the Earth's surface too much to bear, is an extreme being in comparison to our own adaptations – but what determines an extremophily? Is it an evolutionary viewpoint? Does the earliest environment to contain life define what is considered *normal*? If life had arisen in a high-temperature, zero-oxygen hydrothermal vent, or around a cooler alkaline spring at the base of the earliest oceans, would that then be considered as normal, and every environment that has arisen since extreme? Or perhaps extremity is a physical state. All physical factors are on a continuum, and any changes in the conditions that make it difficult for organisms to function are therefore considered extreme. Extremophiles are extreme only in relation to the capabilities of other cells. To a bacterium living in the crushing pressures and high temperatures of an ocean-floor black smoker, we humans living at the level of atmospheric pressure present at the Earth's surface – 1 atmosphere (1atm or 0.101 megapascals/MPa) – breathing oxygen and basking in moderate temperatures, are the extreme beings. Obviously, therefore, normal is relative to whoever is making the comparison, so from a human perspective a normal condition or environment refers to conditions in which humans could comfortably survive without artificial aid.

Must an extremophile actually *love* an extreme environment as its name suggests or can it merely tolerate it? And must an organism depend on these extreme situations for the entirety of its life cycle? An extreme-loving life form does not need to be completely smitten with the conditions it currently resides in. It may be an obligate organism that requires these particular extreme conditions to survive, or facultative – meaning it is not only able to tolerate certain extreme environs when necessary, but is also able to thrive in others at particular stages of its life cycle. The bacterium *Deinococcus radiodurans*, the present gold-medallist of radiation resistance, is widely regarded as an extremophile par excellence, yet retains its radiation superpower only as long as other extreme conditions are lacking; it is severely diminished under freezing or desiccating situations. Other examples are spores, seeds and eggs, which are all far more resistant to environmental extremes compared to their vegetative or animal forms. Similarly, trees, frogs, insects and fish shift their physiology as the seasons change, so they can tolerate remarkably low temperatures during the winter months.

The Extremes

Liquid water will arguably be the cornerstone of any life in our Solar System and beyond. Life also requires an input of energy but crucially must be able to control this energy as it courses through its system. It does this using *redox chemistry*, whereby the loss or gain of electrons between molecules, atoms or ions occurs through a process of oxidation or reduction. Redox chemistry is universal to life on Earth and as all life is based on carbon organic chemistry, it is assumed that such reactions must be allowed to operate for a life form to exist. In terms of extremophiles, they must either live within environments that adhere to these energy parameters or be able to guard against the hostile outside world in order to maintain these conditions within their cells. Life on Earth has been found enduring physical extremes (for example, temperature, radiation or pressure) and geochemical extremes (such as desiccation, salinity, pH, raised or absent oxygen levels or redox potential). It could also be argued that there are also biological extremes such as nutritional availability, and excesses or lack of population density, parasites and prey.

Temperature

More than 80 per cent of our planet, far from being a balmy paradise, exists at temperatures colder than 5°C (41°F) and, even more incredibly, these frigid places are inhabited. As such, the most well-known extremophiles are those that can work to adjust their thermostats, allowing them to live in the coldest, and hottest, places on the planet.

There are two top prizewinners among temperature-tolerant extremophiles. *Thermophiles* are heat-lovers, commonly found in hot springs and geysers the world over. As microorganisms cannot regulate their own temperature as animals do, they must instead adapt all of their cellular machinery to a particular set of operating conditions. Importantly, they have evolved reinforced proteins to hold themselves together against the violent shaking caused by thermal motion at high temperatures. If a protein shakes to such an extent that it loses its three-dimensional structure and becomes denatured, it loses its function and life cannot survive without working proteins. Temperatures approaching 100°C (212°F) normally cause this denaturation to occur in DNA and RNA as well as in proteins, and increases the 'leakiness' of cell membranes to lethal levels. Above 150°C (302°F), many organic molecules decompose entirely. Chlorophyll degrades above 75°C (167°F) preventing photosynthesis from continuing within cyanobacteria and all plants, thereby leading to their death from starvation. The most *hyperthermophilic* organisms (extreme heat-lovers with maximum growth at temperatures over 80°C/176°F) are archaea. Among these, *Pyrolobus fumarii* is a chemolithoautotroph that obtains energy from the oxidation of inorganic compounds, and carbon from the fixation of carbon dioxide; it is capable of growing at temperatures of up to 113°C (235°F). Another archaean, dubbed *Strain 121*, is found growing quite happily at over 120°C (248°F). At these hotter temperatures, the solubility of gases such as oxygen and carbon dioxide decreases, so many hyperthermophilic organisms also anaerobic do not require or use oxygen. The more standard thermophiles are mainly found among the phototrophic bacteria, eubacteria and archaea. The eukaryotes, however, such as algae and fungi, are comparative wimps with an upper temperature limit of only around 60°C (140°F), while for vascular plants it is a pitiful 48°C (118°F), and for fish a relatively frigid 40°C (104°F). It is within this thermophilic group that the universal ancestor of life is commonly thought to have resided (see Chapter 4 for more on LUCA), and may be the type of life found on worlds closer to their stars than the Earth, or around the geysers of the gas giant moons.

The cold-lovers on the other hand, the *psychrophiles*, inhabit environs where the temperatures dip below 0°C (32°F). These communities have enzymes and membranes that are loosened so they can remain dynamic and keep the cells active even at temperatures down to −18°C (−0.4°F). Without this adaptation, as the temperature dropped, the cells' contents would become rigid and inflexible, leaving them unable to fulfil their roles. Temperatures below freezing are a challenge for life as a consequence of the properties of its main solvent, water. Below 0°C (32°F), water freezes into ice crystals that can rip cell membranes apart and without liquid water, solution chemistry within cells stops. The freezing of intracellular water is almost always lethal to life. Despite this, many microbes and cells can be successfully preserved at −196°C (−320.8°F), which is the temperature of liquid nitrogen. In fact, human eggs in fertility clinics are preserved using this although they, together with microbes and cells, are not active at this temperature but held in suspended animation. Among animals, the Himalayan midge incredibly is still active at −18°C (−0.4°F), while the poor Antarctic icefish suffers from heat exhaustion above 4°C (39.2°F). Some of the most extreme psychrophiles reside inside solid icebergs within tiny channels of salty water that are kept liquid owing to their saltiness, and are killed at human body temperature. It is these extreme organisms that can survive the commonly lethal effects of the cold, which are of great interest in the search for life on the icy moons of Jupiter and Saturn.

In organisms more complex than microbes, it is perhaps their behaviour more than their biology that enables them to overcome the physical challenges of extreme temperature environments. As a defence mechanism, they can retreat from unfavourable conditions and relocate to a safer home. In the deserts of Earth, some animals have diurnal habits whereby they bury themselves in the more humid and wet layers beneath the surface to avoid the scorching Sun. In particular, the desert ants of the Sahara are among the most heat-tolerant species in the world and can be found sprinting across the scorching sands. They deliberately come out at the hottest point in the day, when surface temperatures are around 60°C (140°F), which, crucially, restricts their predators' activities. The ants scavenge for the corpses of insects that have died of heat exposure and, although they are physically evolved to resist the high temperatures, could still die rapidly from heat shock themselves. They survive because they stay out for short periods and have long legs, enabling them to move quickly with as little contact with the sand as possible to prevent the heat from building up in their bodies. At the other end of the spectrum, some species of nematode worm in Antarctica can withstand the harsh cold temperatures and lack of water by producing antifreeze and drying themselves out, letting the wind blow them around until water is found again, *i.e.* sitting tight and waiting it out. The red flat bark beetle of northern Alaska is an excellent psychrophile; the formation of ice crystals in its internal fluids is the greatest threat to its survival, but the beetle produces antifreeze proteins that stop water molecules from grouping together. Their larvae have been found surviving at temperatures of $-150°C$ ($-238°F$), for which the antifreeze proteins alone would not be enough. These beetles also deliberately dehydrate their internal tissues; internal water cannot freeze if it is not there any more!

pH

After temperature, the acidity or alkalinity of an environment affects life greatly. Acidity is rated on the pH scale, which measures the concentration of H+ ions (protons) in a solution. Low pH is an acidic environment, while high pH is an alkaline one; pH7 is neutral. Biological processes tend to favour the middle range of the pH spectrum, around pH7, and intracellular and environmental pH often falls in this range as well. Concentrations of protons and their movement from one area to another within a cell are a fundamental mechanism of energy transformation. Proteins commonly denature at exceptionally low pH conditions, yet this is where *acidophiles* are found thriving. Acidophiles are able to survive in highly acidic environments as they can protect the vital molecules inside their cells, such as DNA and proteins, from the high concentration of protons in their environment. These organisms work constantly to pump the excessive levels of protons back across their cell membranes to the outside, like a sailor trying to bail out a leaking ship. Many acidophiles can tolerate pH2, and some as low as pH0. This is best characterised in the red alga *Cyanidium caldarium*, which has been discovered in nature at pHs as low as 0.5, although it grows most successfully in a laboratory at pH2–3. As well as acidophilic prokaryotes, eukaryotic life forms may also be active in environments lower than pH3, although many of these are acid tolerant rather than truly acidophilic and may grow equally well or even better in more neutral habitats. Most eukaryotic acidophiles are, however, still microbial and many yeasts and fungi can grow in acidic soils and peat bogs of pH3–5. A number of filamentous fungi has been found growing at above pH3, and protozoa within acidic, metal-rich waters of pH2–3. Acidophiles such as these could find quite an acceptable home in the acidic cloud decks of Venus.

Alkaliphiles on the other hand prefer a high pH (commonly 8.5–11) and alkaline environments, although they find it equally challenging. As with low pH, there is often a difference of two or more pH units between the internal and external milieu of the cell and alkaliphiles can struggle to generate energy with too few protons in their environment. If cells are to survive in an alkaline environment they must make their own cytoplasm more acidic to buffer the alkalinity and bring it closer to a comfortable neutral value. Representatives of all domains and kingdoms of life are able to tolerate pH as high as around 11, but perhaps the best understood are the alkaliphilic bacteria and archaea, such as *Natronomonas pharaonis*.

Water Availability and Salinity

Water possesses a number of properties such as a high melting and boiling point, a wide temperature range within which it remains a liquid, and it forms hydrogen bonds, which makes it essential for life. It makes up 95–99 per cent of the total molecules in invertebrates and a typical adult human cannot survive the loss of even 14 per cent of his or her water. As such, a lack of this life-giving fluid constitutes a pretty extreme environment. Organisms that can tolerate extreme water loss or desiccation enter anhydrobiosis, a state characterised by little intracellular water and no metabolic activity. A variety of organisms can become anhydrobiotic, including bacteria, yeast, fungi, plants, insects, tardigrades, mycophagous nematodes and the shrimp *Artemia salina*. A further variety of organisms use desiccation-resistant spores to survive dry periods as well as for dispersal, e.g. by the wind. Resurrection plants such as *Craterostigma plantagineum* are unusual in that the plant itself can survive desiccation and can even revive after several months in an air-dried state. Some can even survive a loss of chlorophyll (the pigment required for a plant to be able to photosynthesise). All of these organisms would be very well suited for survival on any planetary body where liquid water is scarce.

Organisms can live in a wide range of salty environments, from essentially pure distilled water to completely saturated salt solutions. The latter, however, is much more problematic for life. Salt-loving *halophiles* grow in high-salt solutions including the Dead Sea, which is not actually all that dead! Very salty or briny water outside the semi-permeable membrane surrounding each cell risks drawing water out from the cell, leading to dehydration. Conversely, if the salt concentration of the environment outside is lower than that inside the cell, it could swell to bursting as water rushes in. This process is called osmosis. Some halophiles have modified their inner workings to cope with higher salt levels, keeping their insides balanced with the outside, and thus safe from osmosis. Others have taken a different approach and packed their cells with different solutes (chemicals dissolved in the fluid inside the cell) to produce an equally concentrated solution to guard against osmosis, while avoiding the issues of a briny interior. These cells protect their innards from becoming too dry or salty by keeping them agreeably sugary. With these adaptations, organisms are able to thrive in the high salt content of salt evaporation ponds at roughly 10 times the concentration of salt in the ocean. Most halophiles are archaeal and bacterial but humans, along with most plants and other vertebrates, cannot tolerate high-salt environments.

Radiation

Radiation, both ultraviolet (UV) and ionising, is particularly hazardous to life. It can damage every single biopolymer (long chains of molecules strung together, produced by living organisms), causing destruction of nucleic acids, proteins and lipids. Radiation is a huge problem beyond the Earth's atmosphere, both within space itself and on worlds themselves lacking a protective atmosphere to shield life from incoming solar and cosmic radiation. As such, any life form that can survive high levels of radiation, or has the ability to recover from radiation damage, has a distinct advantage for survival. Because of the importance of keeping biopolymers intact and functional, organisms can avoid exposure by living underground or producing UV-attenuating pigments. However, because radiation damage cannot always be avoided, there are multiple mechanisms for DNA repair found in all organisms. Still, a few organisms stand out in their ability to handle radiation damage. The bacterium *Deinococcus radiodurans*, as mentioned earlier, is the reigning champion, having been found living within the cores of decommissioned nuclear power stations. It has the ability to withstand both ionising radiation (doses of up to 20 kilograys of gamma radiation) and UV radiation (doses of up to 1,000 joules per square metre) – levels that are 3,000 times higher than what is fatal for humans – but this extraordinary endurance is in fact thought to be a by-product of resistance to extreme desiccation.

Pressure

Life is sensitive to pressure, be it atmospheric (in the air), hydrostatic (underwater) or osmotic (within cells), since any form of pressure forces changes to the volume within cells. Pilots and divers must remain aware that rapid changes in pressure upon ascent can result in gases – generally nitrogen – coming out of solution in the blood to form gas bubbles and, if not treated, this can result in death. A similar situation occurs in microbes from the ocean floor that have gas-filled vacuoles. If they undergo decompression too rapidly, the vacuole expands and bursts, and they die. However, most microbes found in the deep ocean are able to grow at normal atmospheric pressure if decompression occurs gradually. Organisms that can survive in moderately high hydrostatic pressures, greater than the level of atmospheric pressure (1atm/0.101MPa) present at the Earth's surface, are called *piezophiles* – the pressure-lovers, known until recently as barophiles (weight-lovers). The boiling point of water increases with pressure, so water at the bottom of the ocean remains liquid at up to 400°C (752°F). Here piezophilic bacteria are found to grow at pressures up to 500atm (50MPa), and the most extreme piezophilic life at even greater pressures. To survive these high-pressure environments, cells have increased binding capacities of enzymes and extra fatty acids within cell membranes to help them retain their flexibility and motion. Piezophiles include microbes, invertebrates and fish. There is life under high hydrostatic pressure even in the deep trenches of the ocean, living at up to 1,091atm (110.6MPa), for example in the Mariana Trench, the world's deepest sea floor at 10,898m (6.8 miles). *Colwellia MT41* is a psychropiezophilic bacterium, a cold- and pressure-loving polyextremophile, found growing in the deep sea at 1,016atm (103MPa,) at only 8°C (46.4°F). Pressure may be a critical factor for higher animals and plants including ourselves, but neither the highest nor the lowest pressures encountered in the habitable parts of the Earth's surface are obstacles to the establishment of microbial life. Low pressures are too rare on Earth for such communities to have evolved here, but that does not mean it would not be possible on other worlds. Earth's pressure-lovers would be excellent organisms to colonise the base of the oceans on Europa or Enceladus, or to live deep underground on Mars.

Oxygen

The Earth has been anaerobic or oxygen-free throughout most of its history, so our reliance on oxygen is only a very recent phenomenon. Living organisms use energy released by respiration for their life processes and either do this through aerobic (which needs oxygen) and/or anaerobic (which does not) pathways. Today, this need for oxygen is limited to only a handful of life forms on the Earth, so even though to us breathing oxygen is normal, we could be considered the odd ones out – the extremophiles. A huge variety of organisms are found to inhabit strictly anaerobic environments, where the mere presence of oxygen would be toxic. Instead, some bacteria and archaea use elements such as nitrogen or sulphur as the main source of their energy. However, a metabolism using oxygen is far more efficient, although unfortunately this efficiency comes at a price. Molecular oxygen is highly reactive. The reduction of oxygen to water occurring during aerobic respiration in animals, and the reverse during oxygenic photosynthesis in plants, creates hazardous chemically reactive molecules, particularly the hydroxyl radical ($\cdot$OH). Without the superior generation of energy from aerobic over anaerobic respiration, it is unlikely that animals would have arisen owing to their high metabolic demands; cellular damage, or more specifically oxidative damage or stress caused by an excess of these oxygen-rich radicals, is the price we pay. Current thinking suggests that much, if not all, degenerative human disease involves oxidative damage. The need for oxygen for large and energy-intense organisms such as humans nonetheless outweighs these negative effects.

Nature's Superheroes – the Water Bears

The epitome of a polyextremophile and the kings (or queens) of surviving extreme environments, the tardigrades are incredibly endearing, eight-legged, all-but-indestructible and mainly microscopic animals. First named *tardigrada* from the Latin meaning 'slow walker', they are also known as water bears (a name I love, derived from their resemblance to eight-legged pandas) and even moss piglets (drawing comparisons to pygmy rhinoceroses and armadillos). Most tiny invertebrates dart about frantically but the water bears see no need for this; they shuffle along slowly, clambering across bits of debris, ambling around their habitats on pairs of short, stubby legs located under their bodies. The legs are outfitted with a number of hooked claws that resemble the talons of bears.

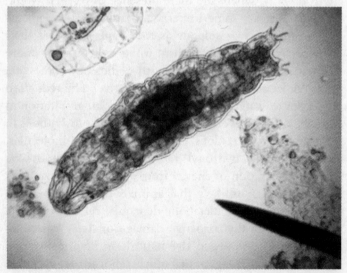

Figure 4 *Light microscope image of a water bear (credit: Louisa Preston).*

Water bears have five body sections, including one that is obviously a head (with or without a pair of eyes) and are encased in a rugged yet flexible cuticle that must be shed as the organism grows. Animals generally grow by adding more cells or by making each cell larger and the water bears for the most part do the latter, as they must must break out of the cuticle in order to grow. All of this houses a nearly translucent, charismatic miniature creature only half a millimeter in length, about the size of the full stop at the end of this sentence.

These mostly microscopic aquatic animals can be seen with the naked eye in the right light and are found just about everywhere across the Earth, from the Arctic to the Equator, from freshwater droplets within garden moss to the salty deep ocean, and to the tops of forest canopies and the summits of mountains. The vast majority of water bears feed only on plant cells or bacteria, slicing them open with their dagger-like teeth and drinking their fluid contents. Others, however, are vicious predators. Moving incredibly fast on the first six legs, they employ their fourth pair to stand upright and attack prey with the rest of their claws – not unlike an actual bear.

The ubiquity of water bears is linked to their best-known feature, their survivorship – they have survived all five mass extinctions – quite possibly because of their strong determination to overcome a cacophony of spectacularly extreme conditions. This has earned them the title of the *most extreme* survivor of all, beating penguins in Antarctica, camels in the desert and the common cockroach. Only the land-dwelling water bears can boast this title however; marine and aquatic species appear to not have developed these superhero characteristics. All the survival adaptations water bears display were selected in response to their rapidly changing terrestrial-based micro-environments. Terrestrial water bears, for example, technically live on the land but actually reside within thin films of water. Moss and lichens, for example, provide sponge-like homes dissected by a myriad of small pockets of water for water bears to inhabit, but are always at risk of drying out. The water bears have two choices in this situation – die or adapt to new, drier environmental conditions. As such, terrestrial water bears have three basic conditions of life: active, anoxybiosis and cryptobiosis. When active they eat, grow, fight, reproduce, move and go about their normal daily routines. Anoxybiosis commences when a water bear finds itself in a low-oxygen environment. Prolonged asphyxia results in failure of the systems that regulate body water, causing the water bear to puff up like a newly-popped piece of popcorn and float around for a few days until it can resume active life. Cryptobiosis effectively resembles death and resurrection. It is a suspension of the water bear's metabolism brought on by the loss of liquid water and extreme desiccation. As its surroundings lose water, the water bear dries up with them, losing up to 97 per cent of its body moisture, shrivelling into a structure about one-third its original size, called a *tun*. In this almost mummified state of anhydrobiosis – meaning life without water – this hardy creature can survive just about anything thrown at it. Water bears actually form tuns several times a year in nature simply by retracting their legs and head and curling into a ball, surrendering nearly all of their body's water. The water bear in effect preserves itself by becoming a powder comprising the ingredients of life, held in suspended animation. When finally rehydrated by an adequate source of moisture, it returns to its active life in as little as a few minutes.

Water bears in their hibernating tun state have been experimentally subjected to temperatures far below freezing (down to −272.95°C/−459.31°F) and, once warmed and rehydrated, returned eagerly to active life. They have been boiled alive, exposed to 150°C (302°F), and still been revived. They have also been weighed down by nearly 400atm (40MPa) of pressure (equivalent to that felt at the base of the ocean) and exposed to superfluous concentrations of lethal gases, such as carbon monoxide, carbon dioxide, nitrogen and sulphur dioxide, and still they returned to life.

How they can survive all of this remains something of a mystery. It may in part be linked to the fact they make excellent travellers. Water bear tuns are almost indistinguishable from dust grains in both appearance and size, and as such can float on the wind in a similar way to spores, pollen and seeds. Just like the latter, the tuns have a preference for where they land and many micro-environments will be unsuitable habitats for freshly arrived water bears. However, an unfortunately placed tun is able to wait for a change in conditions to something more favourable or to be picked up by the wind again and, with any luck, taken somewhere better. When the right watery conditions are finally found, life can begin again. Contributing to this success is the fact that many water bears are able to produce eggs without mating, and in a few cases are hermaphroditic, so able to self-fertilise. A lone water bear may thus be able to establish an entirely new population once it finds the right landing site. It appears they are creatures with few weaknesses; in fact, their only flaw is a vulnerability to mechanical damage when not in their protective tun phase – in other words, you can squash them!

Their near-indestructability may actually be written right into their DNA. Although still undergoing scrupulous investigation, recent sequencing of the water bear genome has revealed that a certain portion may be of foreign origin. Potentially up to 17.5 per cent is made up of a mixture of around 6,000 genes from bacteria, archaea, fungi and even plants, which the water bear has absorbed into itself like a sponge, a process called *horizontal gene transfer*. This is not unusual within bacteria, which trade genes with each other as easily as we might swap emails, but these gene transfers are rarer in animals. A few other examples include ticks that have borrowed antibiotic-making genes from bacteria, aphids that have stolen colour genes from fungi, and wasps that have turned virus genes into biological weapons. One group of genes actually known as the 'Space Invaders' has even repeatedly jumped between multiple organisms including lizards, frogs and rodents. Never has this new alien DNA, however, made up more than one per cent of the new, updated genome. The water bears are quite possibly the remarkable exception. How is this possible? It is thought that when they return to life after drying out in times of low water availability, their cells become sieve-like and molecules from the environment, including potentially any nearby DNA, can enter. Since they are so good at repairing DNA damage, this patching-up ability seals in the new DNA and makes it part of the water bear genome. It has been found that the water bears can even switch on several of their borrowed genes, which in other organisms are involved in coping with stressful environments. Perhaps they owe at least part of their legendary durability to these genetic donations.

Most exciting for astrobiology is the water bear's ability to appear ultimately unaffected by the rigours of space travel; they are the first multicellular animals to outlive exposure to the deadly conditions of the cosmic environment. In 2007, researchers in Europe launched an experiment on the European Space Agency's BIOPAN 6/Foton-M3 mission that exposed tun-state water bears to the solar radiation, heat and vacuum of space, while orbiting the Earth at a distance of 260km (160 miles). When they returned to Earth and were given a little water, the animals soon began to move and feed, and over time grew and reproduced. They had survived an environment in which life as we previously knew it could not. Later in the summer of 2011, Project Biokis-Tardikiss, sponsored by the Italian Space Agency, ferried water bears yet again into space, this time on the US Space Shuttle *Endeavor*. Colonies were exposed to variable levels of apparently lethal ionising radiation but upon return to Earth showed a very high post-flight survival rate, apparently unaffected by the cosmic radiation or the microgravity. The water bears are, for sure, creatures that could survive on any number of worlds in the Solar System so long as evolution was able to progress as it did on Earth to allow them to come into being. Could we one day find similar microscopic animals lurking in pockets of liquid water on Mars or Europa?

Extremophiles from Space?

The theory of panspermia mentioned earlier says that reproductive bodies of living organisms can exist throughout the Universe and develop wherever the environment is favourable. This implies that conditions beneficial for the development of life prevailed at different locations in the Universe and at different times, and may be ultimately responsible for the advent of life on Earth. One of the major criticisms levelled at panspermia, however, is that living organisms could not survive their long journey to Earth, owing to exposure to the nutritional wasteland of space with its solar and galactic radiation, frigid temperatures and vacuum, let alone the fiery descent through the Earth's atmosphere. The Long Duration Exposure Facility (LDEF) and BioPan space experiments sent halophiles into Earth orbit, showing that these salt-loving microbes, as well as water bears, can survive in space. This has led scientists to seriously reconsider the ability of living biological material to travel between celestial bodies, particularly focusing on extremophiles, which are those most likely to survive the trip.

After the water bears, of course, the most probable terrestrial organisms to survive conditions in space are microbes, which might feasibly be stored, protected and transported within comets or meteors. In the vastness of space, microgravity is not lethal to life forms, and the extreme cold and lack of liquid water are survivable by many. Transit times between the expulsion of a rock carrying microbes from its host body to its final destination cannot currently be estimated, however, so we cannot address adequately the nutritional needs of organisms during the journey. We can hypothesise that the exceedingly low metabolic rates resulting from the cosmic extremes of cold and desiccation would render nutritional needs almost non-existent. Thus, we are left with two potential show-stoppers, namely radiation and the space vacuum. Most damage to microbes exposed to space, if they were not protected by a comet or asteroid, would be due to UV radiation, especially in the short term, although heavy ionising radiation has a greater probability of being lethal. Although the data are controversial, *Deinococcus radiodurans*, our extremophilic radiation specialist, did not survive its several-month residency in space and its DNA had extensive unfixable breakages, while *Chroococcidiopsis*, a desiccation-tolerant cyanobacterium that on Earth lives within rocks, survived only 30 minutes when exposed to UV radiation similar to that experienced on Mars. Interestingly, the salt-loving halophiles *Synechococcus* and *Haloarcula-G* were shown to survive for two weeks in space as long as they had some rock or soil shielding them – and could probably last much longer.

The Solar System, and in fact the Universe, is an extremely hostile environment for life as we know it. Even the strongest organisms on the Earth, the extremophiles, would find it a challenge. Yet, knowing these tough organisms exist and have evolved survival strategies to endure within the most extreme environments of the Earth gives astrobiologists renewed hope that something similar, maybe even resembling a water bear itself, may have found a way to thrive in places previously thought inhospitable out in the cosmos. Suddenly the once barren, hostile Solar System has been lit up with biological possibilities.

CHAPTER SEVEN
Searching the Solar System

All members of our inner rocky planetary family – Mercury, Venus, Earth and Mars – are unique and have very distinct differences that have ultimately determined whether or not they might bear witness to life. At the heart of this lie two key attributes: their distance from the Sun, or heliocentric position (which influences their temperature), and their mass. Although celebrated in every mythology throughout history and some even visible to the naked eye on a clear night, their physical characteristics only became known in the twentieth century with the birth of spectroscopy and photography, and subsequently through space exploration itself in the 1960s.

Beyond the orbits of Mars and the asteroid belt, we find the outer gas giants and icy bodies sitting well beyond the Solar System's traditional Goldilocks zone. However, observing these worlds, and especially their moons, has caused us to re-think what actually constitutes a habitable zone for life, how far this zone reaches across the Solar System and, most importantly, whether there could actually be multiple Goldilocks zones located within different planetary families. The opening of our minds to the many extreme conditions life can endure, and all the extraordinary environments in which life on Earth has been found to be present, has meant that the cold, dark outer reaches of the Solar System have begun to intrigue us from an astrobiology perspective and may actually hold the most promising targets for finding alien life.

Messenger to the Gods

Baked in scorching sunlight by day, and deep-frozen by night, it is far too easy to dismiss Mercury in the search for life in the Solar System. Only slightly larger than Earth's moon, Mercury is already the smallest of the Solar System planetary family and is still shrinking. It was long thought to be something of a relic, a stagnant world that had not changed in eons and eons. Like the Moon, it is covered with craters caused by billions of years of bombardment into its iron-rich basaltic crust. This is because Mercury has very little atmosphere to prevent impacts from meteorites and asteroids damaging the surface. The planet is actually home to one of the largest impact basins in the Solar System: the Caloris Basin. Mercury's day-side is super-heated by the Sun to a sweltering 427°C (800°F), but at night temperatures drop hundreds of degrees below freezing. Its egg-shaped orbit takes it around the Sun every 88 days, although this orbit is quite odd – it takes longer for it to rotate on its axis and complete a day than it takes to orbit the Sun and complete a year. It also has only 38 per cent of the gravity of the Earth (a human would weigh 62 per cent less on Mercury than on the Earth) and coincidentally has almost the same gravity as Mars.

We do not know who first discovered Mercury but it has been known to exist since ancient times. *Mariner 10* was the first spacecraft to visit in 1974 and was also the first spacecraft to slingshot past one planet on its way to visit another, as well as the first probe to visit two planets in one mission. Despite being a spacecraft that seemed to be behaving neurotically, posing problem after problem that confounded its designers and controllers, it still managed to reveal a small, bleak planet with a thin helium atmosphere, a weak magnetic field and a cratered surface reminiscent of that of our Moon. The spacecraft ran out of fuel in 1975 and today *Mariner 10* is presumed to be silently continuing its orbit of the Sun. After the picture painted by this probe, it was another 30 years before humanity returned to Mercury – little did we know that the least explored of the inner planets in the Solar System was hiding a very lively personality …

When the spacecraft *MESSENGER* arrived at Mercury in 2011, it went down in history as the first spacecraft ever to orbit the closest planet to the Sun. To arrive at this point, *MESSENGER* soared through the inner Solar System, performing one fly-by of Earth, two fly-bys of Venus, and finally three fly-bys of Mercury. Before its self-destruction on 30th April 2015 with a planned impact into the surface, it was tasked to map Mercury's surface geology, study its magnetic field and expose its internal workings. Apart from creating the most detailed and accurate 3D map of the planet to date, *MESSENGER* also revealed Mercury to be a unique, geologically diverse world. It has active geology: hollows formed when volatile materials – probably sulphur-containing compounds buried beneath Mercury's surface – sublimated (turned straight from solids into gas), causing the terrain to sink by several tens of meters, and volcanic vents measuring up to 25km (15.5 miles) across, which may once have been sources for large volumes of very hot lava.

Owing to its small dimensions, many scientists believed that Mercury's once hot liquid core would have long since cooled and in essence the planet had become a dead, roasted chunk of rock. We now know that it has a massive core for its size, and as such is still partially liquid. We know too that its dynamo is still functioning, creating a weak magnetic field. *MESSENGER* also solved another mystery about Mercury. It bounced laser beams off its surface to remotely collect information about the chemical elements that make up the planet. Incredibly, water ice was discovered to be hiding in regions in permanent shadow near the north pole. Even more excitingly, carbon-containing organic compounds were found, forming a thin layer of very dark organic material covering part of the frozen water. This material, which is somewhat like tar, coal and soot, is believed to be similar to what has been observed on icy bodies in the outer Solar System and in the hearts of comets. Scientists suspect that impacts as well as solar and cosmic radiation are triggering chemical reactions in the organic material, turning it about twice as dark as Mercury's surface.

Although not previously the most logical choice for life, Mercury has shown us that it has a few nooks and crannies with the makings of a habitable environment. To be clear, no one in truth thinks that Mercury has microbes. What it does have, however, is evidence of volcanic activity in its past, a hot liquid core, a functioning magnetic field, a thin atmosphere and gloopy tar-like organic materials. Water ice has been found frozen at its poles that, combined with a hot core, may hint at the possibility of liquid water forming at depths below the surface. Mercury is not in the

Goldilocks zone of our star but is a key witness to the delivery of ingredients for habitability from the outer Solar System to the inner. Even if Mercury is not itself a good candidate on which to look for ancient or current life, the planet may hold clues as to how life got started on Earth. Finding a place in the inner Solar System, where some of the same ingredients that may have led to life on Earth are preserved, is really very exciting. As a result, humanity is going back. Europe's first mission to Mercury, *BepiColombo*, will launch in 2017, arriving at the smallest terrestrial planet in 2024 to carry out the most extensive exploration of Mercury and its potentially habitable environment to date.

Goddess of Love and Beauty – if You Say So ...

With its similar size, mass and composition, Venus's dimensions are very close to those of Earth, hence it is commonly called its twin and is likely to have a still-functioning internal heat source, perhaps from radioactive decay, similar to the Earth's interior. There is, however, one major yet significant difference between these near-identical siblings. Venus's thick atmosphere makes temperatures on the planet hot enough to melt lead, and therefore it is most certainly too hot to sustain life. Blanketed in clouds, the veiled planet once had oceans much like those on Earth, but these evaporated as Venus heated up. Today, Venus is about 100,000 times drier than Earth and is 460°C (860°F) at its surface. Its atmosphere is 96 per cent carbon dioxide with a thick atmospheric pressure of 89atm (9MPa) – a greenhouse planet where no life currently found on the Earth could ever hope to survive.

As one of the first planets to be visited by spacecraft, due to its relatively close proximity to the Earth and it being a necessary waypoint for missions to Mercury, Venus has witnessed and probably rolled her eyes at the many failed attempts to try and visit her. Despite these, however, more than 20 unmanned explorations have been successful, with Venus the target of a cornucopia of Soviet, American and European probes. Spacecraft have performed various fly-bys, orbits and landings, and even balloon probes have been sent to float through its atmosphere. The American *Mariner 5* spacecraft can boast the first successful encounter, coming within 35,000km (21,750 miles) of Venus in 1962. As it flew past, it established that Venus has scarcely any magnetic field and with reasonable accuracy measured the planet's temperature as being up to 316°C (600°F). The Soviet *Venera* fleet achieved notoriety in 1966 when the *Venera 3* space probe crash-landed on Venus, becoming the first spacecraft to reach the surface of another planet, despite not being intact. In 1975, the descent vehicle from the *Venera 9* orbiter was the first probe to take photos (and even black and white video) of the Venusian surface from the surface itself, revealing shadows, no apparent dust in the air, and a variety of rocks up to 40cm (16in) in size that did not appear eroded by wind or water, as would be found on Earth. *Venera 9* also detected clouds that were up to 40km (25 miles) thick, with bases from 30km (18 miles) in altitude; acidic chemicals in the atmosphere, including hydrochloric and hydrofluoric acid, bromine and iodine; surface pressures of about 90atm (9MPa); a surface temperature of 485°C (905°F); and light levels bathing the world that were comparable to those at Earth's mid-latitudes on a cloudy summer's day.

Two notable missions for the purpose of understanding the possibilities of life on Venus are those of the USA's *Magellan* and ESA's *Venus Express*. *Magellan* arrived in 1990 and with its radar mapped 98 per cent of the surface with a resolution of approximately 100m (330ft). The resulting maps are comparable to visible-light photographs of other planets, and are still the most detailed in existence. *Magellan* greatly improved scientific understanding of the geology of Venus: a world covered in volcanic rocks and vast lava plains, lava channels more than 6,000km (3,730 miles) long, fields of small lava domes, and large shield volcanoes. The probe found no signs of Earth-like plate tectonics, although the scarcity of impact craters suggested the surface was relatively young (less than 800 million years old).

Venus Express arrived 16 years later in 2006 and took up a polar orbit, focusing on long-term observation of the Venusian atmosphere from the surface right up to the ionosphere. *Venus Express* confirmed that eons ago, Earth's twin must have had substantial oceans. It observed lightning on Venus happening more frequently than on Earth, and witnessed a colossal double vortex swirling over the planet's south pole. The probe photographed a *night glow*, an eerie radiance in the night-time atmosphere of Venus, seen as the Sun's ultraviolet light hit the atmosphere, and infrared light energy was released by high winds from the swirling of oxygen (O_2), hydroxyl (OH) and nitric oxide (NO) molecules. This was also the first detection of hydroxyl in the atmosphere of any planet other than Earth – important because it is created by a reaction between oxygen and water. A major question in Venus science is whether it is still geologically active today. In 2015, nearly 10 years after the arrival of *Venus Express* at our sister world, tantalising evidence was discovered in a rift zone for hot spots that change in temperature from day to day, and are the best evidence yet for active volcanism on present-day Venus. On Earth, rift zones are the result of fracturing and cracking of the crust and are often associated with the upwelling of magma from below the surface. This process allows hot molten rock to be released through fractures as a lava flow. Although the surface of Venus is thus proving to be still geologically alive and so could provide an energy source for biological reactions, the hellish environment means that life would be hard-pressed to survive.

Despite all this, Venus does sit in a very privileged position at the inner edge of the Solar System's Goldilocks zone. Having undergone a runaway greenhouse effect, the surface has become far too hot for liquid water or organic molecules to be stable, and therefore is not a habitable environment for life as we know it, at least not today. High above the surface, however, is a potential habitable zone where temperatures lie in the range between freezing and 120°C (248°F) and clouds offer long-lasting droplets of water, although they are highly acidic. The lower and middle cloud deck of Venus may, therefore, have the ability to support an aerial biosphere, as hypothesised in Chapter 5; whether it actually does remains a mystery to be solved.

The God of War

The most Earth-like planet known is our reddish dusty sibling, Mars. Half the size of Earth and lacking a magnetic field or thick protective atmosphere, this world is, believe it or not, currently our best hope of finding life elsewhere in the Solar System due to a number of similarities. Its day is only 29 minutes longer than that of the Earth and it takes 1.88 Earth years to orbit the Sun. It has seasons just like on Earth, 38 per cent of our gravity, and has more than five million cubic kilometres (almost 1.2 million cubic miles) of water ice, mostly hidden just below the surface. A toxic atmosphere of 95 per cent carbon dioxide, minimal oxygen and an average temperature of −63°C (−81.4°F), however, makes the surface of Mars appear rather inhospitable to life, especially human life. Astrobiologists are most interested, however, not in the possibility of life existing on Mars today, but in the past. Mars is a planet that we are becoming as intimate with as our own, and so no astrobiological discussion can be started without first acknowledging the many missions that have provided us with this window into such a familiar yet alien world.

Mars Mission History

As we saw in Chapter 1, popular culture had it since the nineteenth century that Mars is or was an inhabited planet, crisscrossed with canals of liquid water built by some advanced civilisation that might or might not be on the verge of colonising the Earth. Once technology caught up with the desire for exploration, however, satellites were tasked to orbit the planet and take the first ever close-ups of the Martian surface. After several catastrophic failures, many of which occurred even before the spacecraft left Earth's atmosphere, in 1965 NASA's *Mariner 4* finally flew by Mars after a 7.5-month journey through 54.6 million km (33,926,870 miles) of open space, snapping the first pictures of the Red Planet. This eagerly anticipated arrival shattered any imaginings of a lush, Earth-like world with flowing rivers and cities full of humanoid Martians. Instead, it was clear that Mars is a rocky, barren world, scarred with impact craters and cavernous valleys – a world that in many ways is more reminiscent of the airless, lifeless Moon than the Earth. It also discovered that Mars has no global magnetic field, which would be necessary to protect any life forms on its surface against dangerous solar winds of charged particles. We now know that Mars's magnetic field disappeared around 4 billion years ago, but we do not know why. With the loss of its magnetic field, the planet's atmosphere was no longer protected and was stripped away, exposing the surface to solar and cosmic radiation, gradually making it even more inhospitable.

Viking

Undeterred by the disappointment of Mariner 4, humanity returned to Mars. In 1976, NASA's *Viking Project* became the first US mission not only to land a spacecraft on the surface of Mars in one piece, but also to return images. Twin spacecraft, both consisting of a paired lander and orbiter, entered Mars's orbit before detaching the landers to begin their fiery descent to the planet's surface. The *Viking 1* lander touched down on the western slope of Chryse Planitia (the Plains of Gold) just north of the equator, while the *Viking 2* lander settled down at Utopia Planitia, the largest recognised impact basin on Mars. Besides taking photographs of a vast number of rocky vistas, the probes became renowned for finding evidence of water action on Mars, including sweeping valleys and deep fluvial erosion patterns. Most famously, however, the landers conducted three biology experiments designed to look for possible signs of life. These discovered unexpected and mysterious chemical activity in the Martian soil, but were unable to provide clear, undeniable evidence for the presence of living microorganisms near the landing sites. The conclusion made was that Mars is self-sterilising: that the solar ultraviolet radiation saturating the surface, the extreme dryness of the soil and the oxidising nature of the soil chemistry combine to prevent the formation of living organisms in the Martian dust. As depressing a result as this seemed to be, scientists now claim that the method by which the samples were collected could actually have destroyed the evidence of life they were looking for ... bad news for *Viking*, good news for astrobiology.

The lack of life found on Mars by *Viking* was an enormous blow to the global community, and it took another 20 years before we successfully went back to Mars. From 1996, *Mars Global Surveyor* (*MGS*), accompanied by *Mars Pathfinder* and its little rover *Sojourner*, orbited and mapped the entire planet. *MGS* achieved so much in its seven-year life, including the characterisation of surface features and geological processes; the determination of the composition, distribution and physical properties of surface minerals, rocks and ice; and the mapping of the global topography, planet shape, and gravitational field. The mission also monitored global weather and, importantly, imaged possible landing sites for the 2007 *Phoenix Lander* and 2011

Curiosity rover. Since then, the space around Mars has become rather full, as NASA's *Mars Odyssey* (2001), ESA's *Mars Express* (2003), NASA's *Mars Reconnaissance Orbiter* (*MRO*, 2005), ISRO's *Mars Orbiter Mission* (*MOM*, 2013) and NASA's *Mars Atmosphere and Volatile Evolution Mission* (*MAVEN,* 2013) have now joined *MGS*, with ESA's *ExoMars Trace Gas Orbiter* (*TGO*), having arrived at the party in 2016. A total of 13 orbiters to date has been sent to circle the planet and map and explore the surface, while several rovers have scoured its landscape searching for clues that might indicate life is, or once was, possible.

MAVEN

The *Mars Atmosphere and Volatile Evolution* mission, or *MAVEN* mission, launched in 2013, is currently orbiting Mars to explore how the Sun may have stripped the planet of most of its atmosphere, turning a world once wet and habitable for microbial life into a cold and barren desert. Scientists want to know what happened to the water that once flowed across the surface and also where the planet's thick atmosphere disappeared to. Each time *MAVEN* orbits Mars, it plunges temporarily into the ionosphere – the ion- and electron-laden atmospheric layer lying uppermost, at 120–480km (75–300 miles) above the planet's surface. This layer serves as a form of shield around the planet, deflecting the intensely hot, high-energy particles of the solar wind. Today, *MAVEN* has shown a plume of atmospheric particles breaking free of the planet's gravity, escaping from the polar region, extending behind Mars like a tail. *MAVEN* has also detected a long-lived layer in the electrically charged ionosphere of Mars, made up of metal ions (iron and magnesium) that are the remains of incoming comet dust and meteorites. The spacecraft has also seen the Red Planet glow under the impact of violent *Coronal Mass Ejections* (*CME*) sent from the Sun. These blast billions of tons of solar material into space at millions of kilometres per hour but because Mars is not protected by a global magnetic field as is the Earth, CME particles directly impact the Martian upper atmosphere, driving the escape of atmospheric gas into space and generating some stunning displays of aurora.

Roving on Mars – *Sojourner*

The great-grandfather of Mars' rovers, *Sojourner* (meaning 'traveller') was the first moving robot on Mars, and indeed the first wheeled vehicle driven on any other planet in the Solar System. It travelled just over 100m (330ft) within the ancient floodplain of Ares Vallis, snapping over 550 photographs of the rocks it encountered. The first one it chemically analysed was dubbed *Barnacle Bill* and this changed our view of Mars forever. The rock was found to have more silica in it than the surrounding environment, a clear sign of past thermal activity. Further rocks, nicknamed *Yogi* and *Scooby Doo*, were recognised as being not just volcanic rock (basalt), but also sedimentary. On Earth, sedimentary rocks are made by deposition of material on the surface and, importantly for Mars, within bodies of water. Images beamed back supported this, exposing rounded pebbles and conglomerates that told a story of rock movement by water in the past. A more water-rich planet was starting to reveal itself, and where once there was water there may have been life.

Spirit **and** *Opportunity*

In 2004, siblings *Spirit* and *Opportunity* bounced on to the surface and finally delivered conclusive proof that liquid water had once been present on Mars. *Spirit* landed in a possible former lake within a giant impact crater given the name of Gusev, while *Opportunity* (fondly known as *Oppy*) headed to the flat plain of Meridiani Planum, where satellite data had found a high level of the mineral haematite, an ore that requires liquid water to form. Their goal was to search for signs of past water activity on the Red Planet and this did not take long at all. As soon as *Oppy* opened its panoramic camera eyes, scientists knew they had struck gold – actually, haematite. Landing in a shallow impact crater, *Oppy* was facing a layered sedimentary rock wall, surrounded by marble-sized iron-rich mineral balls dubbed 'blueberries'. This amazing rover, on its first day of operation, had found an area that had formed in an ancient acidic and oxidising shallow lake. Its mission was already a complete success. The rover had discovered the evidence needed to prove that ancient Mars may have been habitable for life for potentially millions of years. *Spirit* was not letting *Oppy* take all the glory, however, as it also completed its initial mission in record time. At a location in Gusev Crater dubbed 'Home Plate', *Spirit* discovered opaline silica, which would have formed in volcanic fumaroles or hydrothermal vents, showing that water had interacted with magma in the past at that site. It also finally discovered the elusive carbonate rocks, which, given its atmosphere of carbon dioxide and evidence of water, scientists had been expecting to come across far sooner. Equally interesting, *Spirit* also observed complex coatings on olivine basalts, created by modern-day water on Mars, or possibly frost.

Phoenix

Although not a trundling rover, a hugely important astrobiological mission to Mars came in 2008 with the arrival of the *Phoenix* lander. *Phoenix* was designed to study the history of water on Mars and the habitability potential of the Martian Arctic's ice-rich soils. It landed in a flat landscape shaped into 2–3m- (6.5–10ft-) wide polygons, and had been sent there because such geometric features are created on Earth by ice expanding and contracting inside soils when the temperature changes. *Phoenix* found water ice on Mars just a few centimetres below the surface in the middle of the polygons, and amazingly the ice was photographed slowly sublimating when exposed to the Martian atmosphere. *Phoenix* also observed snowfall on Mars, and found calcium carbonate in the soil, indicating a wetter past environment at the landing site. It also located something pretty dangerous for life – perchlorate salts. The big question regarding the presence of organic compounds in the soils surrounding *Phoenix* was left open, since heating of samples containing perchlorate would have caused any organic materials present to break down and be destroyed. Under certain conditions perchlorate can inhibit life, but all is not lost since microorganisms do exist on Earth that can obtain energy from it by anaerobic reduction. Also, the chemical, when mixed with water, can greatly lower the liquid's freezing point, just as salt can when it is applied to roads to melt ice. Thus, although a potential problem for life itself, perchlorate may allow small quantities of liquid water to form on or beneath the surface of Mars today and therefore, ironically, provide microhabitats for life.

Curiosity

Finally, we come to the best-known, and it must be said, most Twitter-savvy rover to date: *Curiosity*. The centrepiece of NASA's Mars Science Laboratory (MSL), this BMW Mini-Cooper-sized robot was detailed with finding out for certain whether Mars is, or was, suitable for life. Its immense size has allowed it to carry a suite of instruments designed to crush, bake and photograph any rock within 2m (6ft) of its robotic arm. Weighing in at 900kg (1,984lb), it can rove up to 200m (650ft) per day, faster than any rover before it (although still literally at a snail's pace), and is powered by a radioisotope thermoelectric generator (or nuclear-powered generator, using plutonium-238). The $2.5-billion MSL spacecraft launched from Cape Canaveral, Florida, on 26th November 2011, and arrived on Mars on 6th August 2012, after a daring landing sequence that NASA dubbed the *seven minutes of terror*. This intricate sequence used a

Figure 5 *Curiosity selfie at 'Big Sky' drilling site taken on Sol 1126. (credit: NASA/JPL-Caltech/MSSS).*

supersonic parachute, rocket thrusters and the now famous skycrane, which allowed the landing assembly to dangle the rover beneath the rockets on a 6m (19.5ft) tether, gently positioning *Curiosity* on to the ground while simultaneously severing the link to crash-land elsewhere on the surface. As *Curiosity* trundles across Mars, it leaves in its tracks a message from home in Morse code. The wheels contain embedded cut-outs of dots and dashes that the rover can use as reference points to estimate how far it has travelled. In dot-dash notation, however, each wheel carries three characters (• – – – / • – – • / • – • •), which just so happens to spell out *JPL*, the acronym of the Jet Propulsion Laboratory in Pasadena, California, which manages the rover mission for NASA.

Curiosity is tasked with searching for habitable environments at the landing site of Gale Crater, a 154km- (95.5-mile-) wide impact crater and central 5.5km (3.5 miles) geologically layered mountain called Mount Sharp, formed by a meteor strike some 3.5–3.8 billion years ago. Scientists chose Gale as the landing site for *Curiosity* because it displays many signs that water was present over its history. Some of these indicators are minerals manifested as clays and sulphates, formed only in the presence of water. They are also exciting to study, as on Earth many preserve biosignatures of past life.

To achieve its goal, *Curiosity* has many instruments and experiments set up on board, including one that bombards the surface with neutrons whose speed would slow if they encountered hydrogen atoms: one of the elements of water. It has a robotic arm that can collect samples from the surface, an oven to bake them inside its main body and the ability to test the gases that are given off, analysing them for clues about how the rocks and soil formed. It has high-resolution cameras that, besides taking great selfies, can take pictures as it moves, providing visual information about the landscape that can be compared to analogue environments on Earth. Panoramic images especially are taken to help scientists select promising future geological targets and to help the rover drivers steer towards those locations to perform on-site scientific investigations.

Within six months of arriving at Mars, *Curiosity* had its answers and the mission was deemed a complete success. Gale Crater had the right ingredients and environments to support ancient microbial life, should it ever have arisen. The landing site itself contains at least one lake that would have provided a deliciously habitable environment. Surprisingly, the clays drilled out from inside some rather informative mudstones at a site known as Yellowknife Bay were found to be much younger than expected. This discovery does not just prove that Mars was habitable but also extends the window of time when it may have been suitable for life. If that were not exciting enough, powder from the very first drill samples *Curiosity* obtained from the surface of Mars included the elements sulphur, nitrogen, hydrogen, oxygen, phosphorus and – you guessed it – carbon! Finally, we have unequivocal evidence that Mars has the chemistry for a habitable environment and the basic elemental building blocks for life.

In December 2014, *Curiosity* went one step further and made the first definitive identification of organics on Mars. It found chlorinated hydrocarbons such as chlorobenzene, dichloroalkane and chloromethane. Also in December 2014 came the news that *Curiosity* had detected wafts of methane in the Martian air. From time to time, Mars belches out a gas that on Earth comes largely from life forms (from one end or the other). This may well hint at communities of microbes living under the Martian surface that are churning out the gas. But in reality, any number of other non-biological, probably more likely, processes can and do make methane. Rocks on Mars contain the mineral olivine, which can react with water to release methane. Also, clathrates or molecular cages harbouring methane in the icy sub-surface could be a source, releasing the gas in bursts over time. Unfortunately, detecting methane alone is not enough to claim life.

In addition to its main mission, *Curiosity* has been carrying out radiation observations to determine how suitable an area like Gale Crater would be for an eventual human mission. *Curiosity* operates its Radiation Assessment Detector for 15 minutes every hour to measure radiation on the ground and in the atmosphere. In December 2013, NASA decreed that the radiation levels were viable for future human crews heading to Mars.

Despite the fact that organics have finally been found on Mars, the question of whether it has or had life still remains. That revelation, and I believe it will come, still lies ahead of us, and will hopefully be provided by the next rover to head to Mars in 2020 – *ExoMars*.

ExoMars

The *Exobiology on Mars* mission (*ExoMars*), set to launch in 2016 and again in 2020, is the first space operation designed chiefly to search for biosignatures of past and present life on Mars. This hotly anticipated astrobiology-led mission is currently under development by the European Space Agency (ESA) in collaboration with the Russian Federal Space Agency (Roscosmos). Launched in early 2016, the *ExoMars* Trace Gas Orbiter (TGO) equipped with an Entry, Descent and Landing Demonstrator Module (EDM) called Schiaparelli, arrived at Mars delivering the lander to the surface on 16th October 2016. Unfortunately, things didn't quite go to plan and Schiaparelli crashed into the surface. It is thought this was due to the lander thinking it was closer to the ground than it was, leading the parachute to release too early and the breaking thrusters to fire for 3 seconds instead of 30 seconds – in reality the lander was still 3.7km above its target. Although it crashed, Schiaparelli still fulfilled its primary aim – to test the landing system for the *ExoMars* 2020 surface platform. TGO on the other hand, at the time of writing, is currently aerobraking to position itself into a 400km (250 mile) high circular orbit above the planet. Once there it will begin its science activities, which include mapping the sources of methane and other atmospheric trace gases. It will also act as a crucial communications relay between the hotly anticipated *ExoMars* 2020 rover and Earth. The unique selling point of this rover is its drill. It will be able to drill 2m (6ft) into the Martian sub-surface, cutting through and exposing millions of years of Martian history, and possibly even buried life forms.

Where would you send *ExoMars* if you had to pick just one spot to visit on Mars? At the time of writing, the choice has just been made. As with every space mission, the decision comes down to a delicate yet familiar balance between engineering constraints and scientific goals. Unfortunately, owing to the way we have to land the rover on Mars the engineers win, and so most of the planet is already ruled out as being too dangerous. The landers need to touch down at as low an elevation as possible to give them more time to come through the atmosphere and therefore more time to slow down (so as not to crash into the surface). To help narrow this already small list of landing choices, a further engineering requirement is for the rover to access as much sunlight as possible since it runs on solar power. *ExoMars* therefore needs to land in a latitude band straddling the equator of Mars that is a meagre 30 degrees wide from top to bottom. Finally, landing on another planet is never possible with pinpoint accuracy; it is much more likely that *ExoMars* will touch down somewhere near its intended target rather than directly on it. This area is called the landing ellipse, and for *ExoMars* is the equivalent of landing anywhere within a 104km by 19km (65 mile by 12 mile) area of where you aimed.

With all these requirements met, the scientific goals start to come into play. Four possible sites were identified for *ExoMars* to make its stand and hopefully achieve its mission to find signs of past or present life – *Mawrth Vallis*, *Oxia Planum*, *Hypanis Vallis* and *Aram Dorsum*. All four sites are found clustered within the same equatorial region of Mars, but covering an area the size of Western Europe, and feature ancient rocks containing a record of the environment on Mars over 3.5 billion years and, of course, display evidence that liquid water once flowed there. Over the coming years the winner, Oxia Planum, will become one of the most studied and talked about spots on Mars.

Mawrth Vallis is the second choice and a veteran landing site, having also made it to the final four for the *Curiosity* rover. Named after the Welsh name for Mars and the Latin for valley, it is an ancient channel carved by catastrophic floods. It has layered cliffs resembling Neapolitan ice cream that are rich in clay minerals. Such minerals, called phyllosilicates, form in the presence of neutral pH water and tell us that habitable conditions for life once existed and, as said before, are also good at preserving signatures of long-dead life. Oxia Planum, 400km (250 miles) away from Mawrth, is also made up of layers of clays and has an ancient channel emptying into a now dry shallow lake. Hypanis Vallis is thought to represent ancient river delta deposits. Here, sediments were built up slowly and may have concentrated the evidence for life, making it easier to find. Finally, Aram Dorsum is an inverted river system, with a hill-like relief instead of a stereotypical depressed river channel. This is quite common on Mars. It happens when water carves a channel and deposits sediments that, once cemented and hardened, survive intact while everything outside is worn away by billions of years of erosion, leaving evidence of an ancient river system, now rising above the landscape instead of below it. Water was flowing throughout this region about 3.8 to 4 billion years ago, a period when life was probably just getting started on Earth, and possibly also on Mars. It may be that evidence of past life on Mars is hiding just beneath the surface at one or all of these sites. At least for now, it looks as though *ExoMars* will be heading to Oxia Planum, with Mawrth Vallis as the back-up, so let us all hope we have made the right decision.

Mars 2020

NASA's *Mars 2020* rover is a mission under development and forms part of a long-term campaign to bring Martian rocks home to Earth. Based on the successful design and operation of *Curiosity*, *Mars 2020* will be sent to investigate an as yet undecided, astrobiologically relevant ancient environment on Mars to uncover its surface geological processes and history, including the assessment of its past habitability and potential for preservation of biosignatures. It has been proposed that the rover collect and package as many as 31 samples of rock cores and soil for a later mission to bring back to Earth. The *Mars 2020* rover will also help pave the way for future human explorers by investigating the ability to use natural resources available on the surface of the Red Planet. Designers of future human expeditions can use this mission to understand the hazards posed by Martian dust and demonstrate technologies to process carbon dioxide from the atmosphere to produce oxygen for human respiration, and potentially as an oxidiser for rocket fuel (see Chapter 10 for more information on humans travelling to Mars).

When on Mars …

Although scientists have not (yet) found life populating the Martian surface, we have not lost hope and are looking harder than ever. Mars has famously undergone global climate change over its 4.5-billion-year life, so we are not just looking to environments on Mars today that might support life, but also at different points in its history. Mars has transitioned through three climatic stages: from relatively wet to semi-arid to hyper-arid conditions, and consequently the surface habitability has deteriorated greatly over its planetary lifetime. The history of Mars can thus be divided into Early, Middle and Present Mars phases. *Early Mars* covers roughly the first billion years of its lifespan, when liquid water was still presumed to be present on the surface. This was an environment similar to the Earth, supporting the hypothesis that life may have had the chance to flourish on Mars as it did on Earth during this time. However, after the first few hundred million years the environmental histories of these two planets diverged drastically. Mars underwent a global climate shift that resulted in a drop in surface temperatures and loss of liquid water. This is the onset of *Middle Mars*, defined as the second billion years when a global cryosphere (during which the entire surface froze) took hold of the planet and paved the way for the Mars we know today. *Present Mars* started some 2.5 billion years ago and is characterised by a hyper-arid climate. The existence of life on the Martian surface today seems unlikely, given the extremely cold and desiccating conditions, high UV radiation flux received on the surface, and the lack of magnetospheric shielding against ionising radiation. Beneath the surface, however, partially protected from radiation by rocks and dust, and where it is slightly warmer so that isolated pockets of liquid water may remain, who knows what (or who) might be hiding ...

King of the Planets

The terrestrial planets do not present many especially habitable conditions on their surfaces today. That said, they are a much better bet than the planets beyond the asteroid belt. Once we cross this rotating barrier of rock, we enter the realm of the gas and ice giants, such as Jupiter. While there have been no samples taken that could test for microscopic life on Jupiter, there is considerable and compelling evidence against life as we know it existing or ever having existed there. Composed mainly of hydrogen and helium, there is virtually no water present that could support a life form. The planet does not have a solid surface for life to develop upon and so the only real (tiny) possibility of finding it would be in floating microscopic form high up in the atmosphere. However, this too has its problems. The atmosphere of Jupiter is in constant chaos, so even if life somehow held on near the lower pressure regions in the upper reaches, and could resist the harsh solar radiation found there, it would eventually be sucked down into realms where there is 1,000 times Earth's atmospheric pressure and temperatures over 10,000°C (18,000°F); it would be almost instantaneously destroyed. No life on Earth could survive in anything close to these environments.

Known, therefore, as completely inhospitable, the gas giants are not truly included in the search for life, although it is fascinating to think about ways in which it might work. The most intriguing astrobiological targets in the Solar System are actually found on the moons orbiting these gaseous behemoths. The first is Europa, which on the face of it does not look or sound particularly appealing to life as it is constantly bombarded with ionising radiation owing to its location within Jupiter's magnetosphere. Temperatures at its surface range from −187°C to −141°C (−304.6°F to −221.8°F), far below the lowest limits for microbial growth − not surprising since it is an average 805 million km (500 million miles) from the Sun. In 1979, the *Voyager 2* probe whizzed past and spotted a network of cracks on Europa's surface, confirming earlier theories that the moon was coated in a thick shell of ice. When the Galileo probe showed up in the 1990s, it became clear that the cracks were occurring because the ice was moving, floating on top of a hidden layer of liquid encircling the moon. Beneath Europa's estimated 100km- (62-mile-) thick icy crust we believe there resides a liquid ocean with more water than covers the surface of the Earth. Within the cracks and fractures of the ice is a salty dark material, quite possibly the same as regular table salt, sodium chloride, which has risen up from the ocean beneath.

Europa is a distant but tantalising world for astrobiologists. The surface ice itself is not an environment that any currently known terrestrial life could withstand, so it's unlikely we will see microbial igloos popping up there any time soon. However, the ice could provide just enough protection from the intense bombardment of radiation and encourage more favourable temperatures beneath it, to allow for the preservation of organics and even life forms. Just as a layer of ice over a pond allows the water beneath it to stay liquid and aquatic life to go on living through a freezing winter, Europa's rind of ice shields its enormous ocean and helps to keep it warm enough to remain fluid in spite of the moon's great distance from the Sun. Yet as Europa orbits Jupiter, the moon is contorted by the giant planet's gravitational field, generating an interior heat, by far the dominant heat source, which also keeps its water from freezing altogether. Potentially, active volcanoes and vents may exist at the base of the ocean, further heating the water, and providing sites where bacterial life may congregate, as it does on Earth. Plumes of water vapour have also been observed by NASA's Hubble Space Telescope in 2014 and again in 2016 erupting from near its equator. As such Europa has many features thought to be key for the development and even persistence of life, such as water and heat energy; astrobiologists are now keen to detect the presence of organic chemicals.

The much-awaited *Jupiter Icy Moon Explorer* (*JUICE*) mission is an ESA spacecraft planned as part of the Cosmic Vision science programme, scheduled to pay a visit to the Jovian system in 2030, to study Ganymede, Callisto and Europa (Io will be left out this time). Hopefully launching in 2022 and taking eight years to reach the system, its aim will be to analyse the character of these three worlds and evaluate their potential to support life, as all are thought to have significant bodies of liquid water beneath their surfaces. In particular, the focus on Europa will be on the chemistry essential for life, including organic molecules, and on the non-ice material criss-crossing its surface.

Lord of the Rings

If life is near impossible on Jupiter, you can guarantee the same can be said for Saturn. Comprised almost entirely of hydrogen and helium, with only trace amounts of water ice in its lower cloud deck, it has no surface upon which life could live. At the top of the clouds, the temperatures are around −150°C (−238°F), and although it gets warmer as you descend through the atmosphere, the pressures increase too. Sadly, once temperatures are warm enough to have liquid water, the pressures are simply too high for life. It is also extremely windy up there, with speeds of up to 500m/s (1,640ft/s). As with the Jovian system, the quest to find life near Saturn is turning its focus away from the planet and towards the moons.

Titan, Saturn's largest, haziest moon, is getting scientists really rather excited. Deceptively Earth-like, Titan has a dense nitrogen-rich atmosphere (the only moon known to do so), complete with clouds and seasonal rainstorms that soak the surface. Sunlight and electrons stream across the moon from Saturn's magnetosphere and break apart the nitrogen and methane in its atmosphere, setting off a cascade of reactions that produce organic compounds, and creates a solid organic haze that fills the atmosphere and shrouds the surface from view. It has a very familiar landscape beneath this seemingly impenetrable veil, with mountains, dunes, riverbeds, shorelines and seas. Indeed, Titan is the only place in the Solar System, besides the Earth, that has liquids pooling and flowing across its surface. It is likely that Titan would be a promising place to look for extraterrestrial life in the Solar System, if not for its coldness; Titan is far too chilly for life as we know it. All water on Titan is found as rock-hard ice. In fact, the many rocks that litter the moon's surface are not made of rock at all but actually water. At Titanian surface temperatures (−179°C/−290°F), phospholipids − the chemical compounds that provide structure to cell membranes − cellular water bodies would be frozen solid. Any life that evolved on Titan's surface would need to be made of a very different set of chemicals and not be reliant on water, as it is locked in a state inaccessible to it. But potentially lucky for life, Titan's puddles are filled with hydrocarbons. Methane and ethane, which on Earth are gasses, are able to flow as liquids across the surface owing to Titan's frigid environment. The volume of liquid hydrocarbons resting in Titan's second largest sea, Ligeia Mare, is actually 100 times greater than all the oil and gas reserves on

Earth combined. Could a bizarre non–Earth-like life form exist on Titan that uses these slick, liquid hydrocarbons in a similar way to how life on Earth uses water?

During the 1990s, the Hubble Space Telescope (HST) offered hints that Titan was a wet world, but this was not confirmed until the NASA-managed *Cassini* mission allowed scientists to get a good look at the moon. This was a collaborative mission that included 16 European countries together with the US. On 14th January 2005, after a seven-year voyage, the *Cassini* spacecraft sent the Huygens probe parachuting through the haze to a spot on Titan's equator, to become the first terrestrial robot ever to land in the outer Solar System. It then sent transmissions from the surface for another 70 minutes before *Cassini* moved out of range. *Cassini* is the fourth space probe to visit Saturn but the first to enter orbit and continued to send data until its demise in September 2017. It has revealed a world on Titan that looks very much like ours – but with a completely different chemistry. As well as a hydrocarbon-drenched realm, Titan also may have a deep sub-surface ocean similar to that within Europa and another of Saturn's moons, Enceladus. It may prove to be a water-ammonia mixture, which could be an environment habitable for organisms with biochemistry similar to that of terrestrial life – although requiring them to power a metabolism at the temperatures present on Titan would be a real challenge, even if the chemistry were usable. At this point, we do not really know what kind of life might be able to survive on Titan, as we have no examples of similar life on Earth. But this does not mean it isn't possible.

Saturn's sixth-largest moon, Enceladus, discovered by William Herschel in 1789, is a tenth the size of Titan, and is covered by fresh, clean ice with a surface temperature at noon of −198°C (−324.4°F). When Cassini flew by in 2005, it drew renewed interest from astrobiologists with the sighting of present-day geological activity occurring at its surface. Jets of fine icy particles and water vapour were observed erupting from cryovolcanoes at the south pole.

Over 100 of these jets have been seen so far, feeding into a large plume that soars several thousand kilometres into space and containing not only water vapour but also simple organic compounds and volatiles – such as nitrogen (N_2), carbon dioxide (CO_2) and methane (CH_4) – similar to the chemical make-up of comets. Some of the water vapour actually falls back on to Enceladus as snow, while the rest escapes and supplies most of the material making up Saturn's E-ring. The southern polar terrain surrounding the source regions of the plume is surprisingly warm considering it is made of ice, and analysis of icy particles within Enceladus' plume strongly suggests the presence of a salty sub-surface alkaline ocean. Most models regarding the origin of this plume include a sub-surface liquid water aquifer, and it is this aquifer with its potential to support the origin and evolution of life that is of particular interest for habitability. A plausible sub-surface ecosystem on Enceladus would be unlike many terrestrial biomes, as life forms would have to be independent of oxygen and not rely on organic materials produced by photosynthesis.

Gods of the Sky and Sea

To sustain life on Uranus or Neptune, these distant planets would need a source of energy that even the simplest life could exploit, as well as some type of standing liquid water. A sister ice giant to Uranus, the surface of Neptune dips to a glacial −218°C (−360°F), while the cloud tops are −224°C (−371°F) in temperature. They are both far too cold to host bodies of liquid water and have no solid surface on which they could form in any case. Uranus is composed mostly of methane, water and ammonia ices enshrouded by an atmosphere of hydrogen and helium; it is methane that gives it its blue-green colour. Tremendous pressures inside Uranus created by the overbearing atmosphere raise the planet's temperature to more than 4,700°C (8,492°F), and would instantly crush and burn life. Add to this the lack of sunlight and internal heat and there is an absence of essential energy for life. Even though it seems impossible, technically there remains a chance some bizarre incarnation of life might be able to survive on Uranus but we are unlikely ever to be able to send a spacecraft down into the planet to check. *Voyager 2* is the only spacecraft to have flown by Uranus, back in 1986. The planet revealed few secrets but there were hints that there exists an ocean of boiling water some 800km (500 miles) below the cloud tops. The five natural satellites in orbit close to Uranus, such as Titania, would be more likely to contain habitable niches, but currently are not deemed to do so and humanity is in no desperate hurry to explore further in the Uranus Planetary System.

Neptune similarly offers little hope of life. It is a cold and dark world, whipped into a frenzy by supersonic winds. About 4.5 billion km (2.8 billion miles) from the Sun, it is mostly composed of a very dense atmosphere of hydrogen and helium, with ices of water, ammonia (NH_3) and methane (CH_4) over a possibly heavier, approximately Earth-sized, solid core. As is the case for the appearance of Uranus, Neptune's blue colour is also the result of methane in the atmosphere. There is very little water in the cloud tops, but the percentage increases as you descend towards the core. Perhaps there is a band on Neptune where there is enough pressure and temperature for liquid water to form into an ocean layer. The only spacecraft ever to have visited Neptune is the same one that flew by Uranus – *Voyager 2* – passing by Neptune three years later in 1989.

A slight glimmer of hope may exist within Triton, Neptune's largest and backward-orbiting moon. It is tremendously cold with temperatures on its surface of water ice of about -235°C (-391°F). Its unusual orbit, essentially heading the wrong way round, implies that it did not form around Neptune but was captured after being ejected from the Kuiper Belt. Triton is quite dense, suggesting that, unlike its parent planet, it may have a solid core of silicate rock. In spite of its frigid state, *Voyager 2* found geysers belching icy matter into space for over 8km (5 miles). This, as with other moons in this distant neighbourhood of the Solar System, could imply that a liquid ocean is hiding beneath an icy crust, kept fluid by tidal friction and the decay of radioactive isotopes, as happens on the Earth. Similarly recorded by *Voyager 2*, Triton's sparse atmosphere has also now been detected from Earth and is actually growing warmer – we do not yet understand why.

The increasing number of sub-surface oceans on icy Solar System bodies could provide potential habitats for primitive extraterrestrial life forms, yet astrobiologists do not realistically expect to find these inhabiting either Neptune or Triton. If the ammonia that may well be present in Triton's subsurface ocean were able to lower the freezing point of water, however, it might be a more suitable host for life. There is nothing to say that life (but not as we know it, Jim) could not be thriving on either body, just waiting to be discovered.

King of the Underworld – the Underdog

Now that the *New Horizons* spacecraft has completed the first Pluto fly-by (in 2015) after a nine-year journey, we have finally visited every member of the Solar System and the secrets of this dwarf planet too are starting to be revealed. But could this once-upon-a-time planet have the theoretical potential to support life when it is more than 4.8 billion km (3 billion miles) away from the warming embrace of the Sun?

It is amazing what, until 2015, we didn't know about Pluto. For starters it is larger than we first thought. About two-thirds the size of the Moon, it is 2,370km (1,470 miles) across and could comfortably fit inside the area of Russia. The fact that we were wrong about something as simple as its dimensions demonstrates the importance of visiting a world to get accurate information, but is also significant in showing that Pluto is less dense than we thought – it turns out to be made of more ice than rock. Ancient surfaces, like those on the Moon, record the history of impacts – and therefore the history of the Solar System – in the form of craters. Pluto, it was assumed, would also be covered in a very old cratered crust but, amazingly, it has areas that are smooth and unscarred by impact craters and composed of much newer icy deposits as seen in the heart-shaped region of Tombaugh Regio. This tells scientists that the surface can only be about 100 million years old – fairly young in geological terms. How is this possible? To smooth away the craters created during Pluto's history, as they were on most of the planets and moons in the Solar System, Pluto would need to have some kind of internal heat to soften or melt the icy surface. We have no idea what this source of warmth might be. It is probably too small to generate much radioactive heat inside its body, and there is no larger parent world to squeeze it and generate tidal energy; yet it is obviously geologically active. Figuring out this conundrum will be a huge revelation for planetary science and astrobiology alike.

Pluto is also losing its atmosphere (yes, surprisingly it has one) and as such has merely 1/100,000th of the atmospheric pressure at sea level on Earth. It has 3.5km- (2-mile-) high mountains made of rock-solid water ice, a frozen copy of the Earth's Rocky Mountains, and a surface that in appearance resembles boiling milk. The smooth plains of Tombaugh Regio have officially been titled *Sputnik Planum* after the first Russian satellite, launched in 1957, an ice sheet within these plains appears to have flowed in a similar way to glaciers on Earth and may actually still be flowing. There are surface patterns resembling the convection cells seen in steadily boiling milk (yes, another milk analogy). One interesting feature is that Pluto has a tail rather like that of a comet, as it is losing an estimated 500 tonnes (550 US tons) of nitrogen into space every hour. *New Horizons* flew through this nitrogen tail, which extends for

109,000km (67,730 miles) away from Pluto, and is sculpted by electrically charged particles that have travelled all the way from the Sun and are continuing past.

Unsurprisingly, there are no signs of life on Pluto, yet it is showing hints of the ingredients we normally use to describe a habitable environment. There is an as yet unknown heat source, there is water – albeit frozen as ice – and the most exciting find, organic carbon-based molecules. Pluto's tenuous atmosphere has haze layers where methane molecules (CH_4) are broken apart by the Sun's UV radiation. These recombine in various ways to form larger, more impressive molecules, but eventually group into solid specks called *tholins*. These have at best been described by Sarah Hörst, an Assistant Professor in the Department of Earth and Planetary Sciences at Johns Hopkins University, as 'abiotic complex brown organic gunk'. These fall as tar-like rain on to the surface, giving Pluto its surprisingly Mars-like reddish-brown colour. With flowing ice, exotic organic surface chemistry, mountain ranges and a vast carbon-rich haze, Pluto is showing a diversity of planetary geology and even astrobiology that is truly thrilling and highly unexpected this far out in the Solar System – Pluto's payback for being demoted to dwarf status!

Comets

In the quest to find evidence of life elsewhere in the Solar System, comets have been implicated in a number of stories, from the extreme notion of transporting living cells or even fully formed microbes from planet to planet, seeding each new world with life, to the more plausible idea of icy rocks carrying the basic organic building blocks to Earth, which contributed to the origination of life.

It is widely known that there are organic carbon-based molecules in interstellar space, with large quantities trapped in interstellar clouds and comets. When a European

spacecraft analysed dust particles from Halley's Comet in
1986, it turned out to be some of the most organic-rich
material ever measured in the Solar System. We now know
for sure that comets could have provided the raw ingredients
that the Earth would have needed for life. A regular visitor
to the inner Solar System, 67P/Churyumov-Gerasimenko
was the chosen target for the ESA catch-a-comet *Rosetta*
mission. It has a short orbital period of 6.45 years, controlled
by Jupiter's gravity, and is believed to have originated from
the Kuiper Belt. When these Jupiter-family comets cross the
orbit of Jupiter, they gravitationally interact with the
massive planet. Their orbits gradually change as a result of
these interactions until they are eventually thrown out of
the Solar System or collide with a planet or the Sun.

In November 2014, everyone's favourite comet lander,
Philae, hopped, skipped and jumped its way into history.
Instead of the planned single landing, *Philae* had an initial
bouncing touchdown followed by a collision with a crater
rim and two further touchdowns. Important for
astrobiologists, analysis of data from the UK-led instrument
Ptolemy discovered molecules that can form sugars and
amino acids. Ptolemy sampled ambient gas and detected
the main components of the coma gases (those in the halo
around the nucleus of the comet), including water vapour,
carbon monoxide and carbon dioxide. Smaller quantities of
carbon-bearing organic compounds were also identified,
such as formaldehyde and acetone. Formaldehyde is
implicated in the formation of ribose, which ultimately
features in molecules such as DNA, and acetone is best
known as the chief ingredient in nail polish remover (both
equally important creations). While this is a long, long way
from finding life itself, Philae has shown that the organic
compounds that eventually translated into organisms here
on Earth were present in the early Solar System and within
moving bodies that could have transported them to the
newly formed planets. So far there are no signs in 67P of
amino acids, the building blocks of proteins. However,

they are probably there somewhere, since they appeared in samples from NASA's *Stardust* mission, which returned material to Earth from the tail of comet Wild 2 (81P/Wild) in 2004, and have also been traced in meteorites that crash-landed on Earth.

There are no life forms as yet identified within comets but they act as couriers, delivering water and organic-rich dust throughout the Solar System – sowing the ingredients for life far and wide. The challenge now is to discover where else they may have ended up.

Every day, our understanding of the envelope of life expands and with it the possible places in which life might exist in the Solar System. Not just that, but after a century of certainty the Solar System itself can still surprise us. At the time of writing, two astronomers from the California Institute of Technology have found evidence of a possible ninth planet lurking in the farthest reaches of our planetary neighbourhood. *Planet Nine*, as it has been dubbed, has not actually been seen yet but its existence has been inferred. Should this mysterious world existing as far away as 1200 AU be proven to be in fact real, who knows what it might be like or if it might even have habitable environments. One day we could find the envelope of life stretched to even greater extremes. Nearly every planet and moon observed so far has the potential to support a habitable environment (not that this proves they ever have had or currently host life), so future studies of these worlds will be extremely exciting. But why stop at the edges of the Solar System when there is an entire galaxy of potentially habitable worlds beyond ...

CHAPTER EIGHT

Extraterrestrial Worlds: Life Not As We Know It

The Milky Way contains over 100 billion stars, and the whole Universe is made up of more than 100 billion galaxies. Surely there is at least one planet out there, just teeming with life, orbiting around one of those stars? We know that other planets exist in other solar systems. On 6th January 2015, NASA announced the Kepler Space Telescope had discovered its 1,000th *exoplanet* – a planet orbiting another star in place of our Sun. In fact, as of October 2017, there have been 3671 planets in 2751 planetary systems found. Nonetheless, of all the worlds discovered to date, only a handful closely resembles the Earth. Instead, they exhibit a truly spectacular diversity, varying immensely in their orbits, sizes and compositions, and have been seen circling a wide variety of stars, including ones significantly smaller and fainter than our Sun. We are starting to see that any kind of world imaginable (within the realms of physics, of course) might be possible somewhere out there, and if that were true then theoretically a huge variety of alien life is possible too. The search for ET often focuses on planets that resemble Earth, the only world known by humans to host life − but does this always need to be the case?

Alien Worlds

Only in the past two decades have astronomers been able to confirm the existence of thousands of worlds orbiting distant stars, and could finally then legitimately question the possibility that some of these exoplanets might be home to extraterrestrial life. However, the conviction and belief that planets outside our Solar System exist goes back long before this. The Catholic monk *Giordano Bruno* proposed in 1584 the notion of 'countless suns and countless earths all rotating around their suns'. And yet, even in Bruno's time, the concept of a plurality of worlds was not entirely novel as the scientists and philosophers of ancient Greece had also pondered whether other solar systems might exist and whether some would harbour other forms of life. More recently, the astronomer *Edwin Hubble* (1889–1953) used the world's most powerful telescopes of the day at the observatory on Mount Wilson in California and by 1923 had established that tiny nebulae visible in the heavens were in fact fields of hundreds of billions of stars, located way beyond the Milky Way. Hubble's observations proved the existence of countless potential worlds out in the darkness of space, and any number of these could be a habitable planet bustling with life.

Too far off from our own planet for direct observation, exoplanets can only be detected through their effects on their host star. Since our Solar System provides an excellent example of a planetary system with life, not surprisingly astronomers began the search for new worlds by examining stars similar to our Sun. Ironically, however, the first genuine discovery of a planet beyond our system came in 1994, when two or three planet-sized objects were found orbiting a pulsar – a dense, rapidly spinning corpse of a supernova explosion – rather than the expected Sun-like star. Although the existence of this small group of planets remains controversial, there is a consensus that these worlds could not support life as we know it, being permanently doused in high-energy radiation. The first discovery of a planet orbiting a star similar to our Sun came in 1995. A Swiss team proclaimed their finding of a new planet at least half the mass of Jupiter, set in a speedy orbit near the star 51 Pegasi. Thus began a surge of exoplanetary discoveries and by the arrival of the twenty-first century, several dozen more worlds had been detected.

Planet Hunters

In the beginning, there was *Hubble*. This first-generation space telescope, launched in 1990, has provided some of the most breathtaking images ever taken of our cosmos, and has been celebrated as the first to take an image of an exoplanet, *Fomalhaut b*. Launched in 2003, the *Spitzer Space Telescope* observes objects in the infrared spectrum and was the first instrument directly to detect light coming from an exoplanet. The data it has collected has revealed the composition, temperature and possible wind patterns on many distant extrasolar worlds. From its launch in 2006, the French *CoRoT* (*Convection, Rotation and planetary Transits*) mission, was the first exoplanet-hunting space operation, looking specifically for signs of planets transiting in front of their local star. It was a major contributor to the list of confirmed exoplanets, including some of the best-studied planets beyond our Solar System. CoRoT ceased to function in 2012, and was retired in 2013.

The $600m Kepler mission was launched in March 2009 and its primary mission came to a premature demise in May 2013. Its mission objective to establish how frequently Earth-like planets, in or near the habitable zone of their host star, occur across the Milky Way galaxy was a hunt for Goldilocks planets. It used a specially designed telescope called a photometer (light meter), that continuously records the brightness of stars. To make its discoveries, Kepler targeted a dense field of stars, allowing it to monitor simultaneously and unceasingly 150,000 balls of burning light. As a planet drifts across the face of its star in *transit*, it blocks a percentage of the light as viewed by the observer, and it is these dips in brightness that enable the planet to be detected. The drops can be miniscule, often around 0.01 per cent for an Earth-sized world, and last between 1–16 hours. The change must occur in a regular sequence to be attributed to a planet orbiting the star. Errors are quite possible at this subtle level, so once a candidate has been identified, Earth-based observatories take over, searching for telltale fingerprints of the host star's wobble as it responds to the pull of the orbiting planet's gravity. This is an invaluable double-check and shows that some one in ten of Kepler's candidates are false alarms. Even though its original mission has ended, it continues to observe the heavens, and scientists and the public alike will be combing through the massive treasure trove of publicly available data for years to come. This is how new exoplanets continue to tumble out of the sky and the number of potentially habitable exoplanets found is still climbing, long after the mission to detect them has ended. Kepler has discovered more than half of all known exoplanets to date, with over 2,000 confirmed and at least another 3,000 unconfirmed candidates.

Owing to its resounding success, new space telescopes capable of finding Earth-sized worlds around nearby stars are being designed to succeed Kepler. At the time of writing, the European Space Agency is scheduled to launch *CHEOPS* (CHaracterising ExOPlanet Satellite) near the end of 2018, and NASA will launch *TESS* (Transiting Exoplanet Survey Satellite) a few months earlier in March 2018. By 2024, ESA hopes to have followed CHEOPS with a larger planet-finder dubbed *PLATO* (PLAnetary Transits and Oscillations of stars). This mission's objective is to identify and study a large number of extrasolar planetary systems, with the emphasis placed on finding Earth's twins. If all goes to plan, the *European Extremely Large Telescope* (E-ELT), currently being built in Chile, will then be able to analyse the atmospheric composition of these newly found planets. By analysing the mix of gases in an atmosphere, E-ELT will be able to determine whether the planet in question is potentially habitable – or even inhabited. In addition to all these missions, the shiny new *James Webb Space Telescope* (JWST) is a sophisticated new observatory currently under design that is tasked with unlocking some of the greatest mysteries of the Universe, and which could also play a key role in the hunt for alien planets. Sold to the public as a replacement for the Hubble Space Telescope, this $8.8-billion infrared telescope is also planned for launch in 2018 (a busy year) and will orbit 1,496,690km (930,000 miles) from Earth, in a region called the *Lagrange Point 2*. Here, the gravitational forces from the Earth and the Sun essentially cancel one another out, so JWST will be able to maintain a stable orbit while using minimum energy. From this orbital perch, it will be able to stare uninterrupted at stars through its sensitive infrared eyes and allow astronomers to 'sniff' the atmospheres of alien planets to break down their molecular composition.

Habitable Worlds

Everyone wants to find a planet that might have life on it, either now or in the past, whether this is a second Earth or a world with sufficiently similar features and conditions that could make it habitable for life as we know it. An Earth-twin would have an Earth Similarity Index (ESI) of 1.0, and all exoplanets get assigned their own ESI. Astronomers remain committed to the idea that planets and moons with liquid surface water are the best bet for finding life, so the goal is to find a Goldilocks planet or moon. With this in mind, they are searching for one that is neither too hot nor too cold nor too large nor too small, but *just right* for liquid water. We know why stellar distance is important, but size also matters. Too small a planet or moon would not be able to hold on to a protective atmosphere, while if it were too huge, it would have an immense atmosphere of hydrogen and helium, making the surface too hot to support life. This holy grail of a perfect world does not need to be a complete replica of the Earth, but should simply enjoy some of its finest features. In fact, the more exoplanetary worlds we see, the more we are starting to consider that the Earth may not even be the gold standard for life. There may conceivably be superhabitable worlds out there that are even better suited than the Earth to support life.

A Periodic Table of Exoplanets divides most of the known candidates into groupings based on mass or dimensions and temperature. Exoplanets in the *Hot Zone* are too close to their parent star to have liquid water, whereas those in the *Warm Habitable Zone* are at the right distance for liquid water to be stable, given that they are the right size (from half that of the Earth up to 10 Earth masses). Water will only be in existence as ice for those in the *Cold Zone*. The *Mercurians* are low-mass bodies, most likely spherical and lacking an atmosphere, like Mercury and the Moon. The *Subterrans* are comparable to Mars, *Terrans* to Earth and Venus, while *Superterrans*, or *Super-Earths*, are up to 10 times as massive as Earth, a category with no comparable examples in the Solar System. *Neptunians* are similar in mass to Neptune and Uranus (are you starting to see a naming pattern yet?), and *Jovians* are compared to Jupiter and Saturn-sized worlds, or greater. The two largest types of exoplanet are, unsurprisingly, the most commonly detected as they are the easiest to spot. These *Hot Jupiters* have similar characteristics to Jupiter itself but generally orbit closer to their parent star and so have far hotter temperatures. *Giant Neptunes* are large gaseous planets, considerably more massive than the Earth but smaller than Jupiter. Both of these types of world are believed to be inhospitable to life.

Searching for Super-Earths

Instead of hunting solely for Earth's twin (in terms of both the planet and its star), we are fascinated by planets similar to, yet larger than Earth – the so-called Super-Earths. This name denotes the size of the worlds, not their capabilities. These are worlds that feel both familiar and yet completely alien. They may be made of rock and metal, or ice and gas. These planets may have oceans and atmospheres, or contain nothing but hydrogen and helium. The goal, of course, in studying these is to find a rocky Super-Earth located in the Goldilocks zone of its parent star. More than 30 Super-Earths had been discovered up to 2016, with the first found around a pulsar in 1992. The first one discovered around a main-sequence star, the red dwarf *Gliese 876*, was not observed until 2005. The first discovery of a potentially habitable super-Earth was in 2007, this time around *Gliese 581* and on the edge of that star's Goldilocks zone. In fact, not one but two worlds were discovered, with *Gliese 581c* getting the most attention. This planet has a mass of at least five Earths and sits on the overly warm side of the Goldilocks zone (conversely, *Gliese 581d* sits on the cold side). Scientists believe that 581c suffered a runaway greenhouse effect, as is thought to have occurred on Venus.

At the time of writing, the most Earth-like planet yet has been discovered and named the ever-imaginative *Kepler-452b*. It is the first almost Earth-sized planet to be found in the Goldilocks zone of a star very similar to our Sun. It orbits a star known as *Kepler-452*, located in the Milky Way and some 1,400 light years away in the direction of the Cygnus constellation. Its size – 1.6 times that of Earth – hints at it being a rocky world that is likely to have an atmosphere, good cloud cover and possibly active volcanoes. This is our best candidate yet for *Earth 2.0* and could also be the ideal place to look for evidence of extraterrestrial life. What is most fascinating about it, however, is its age: it is a staggering 1.5 billion years older than our own Earth. This world may give us a glimpse of what awaits our planet in the future. We have also found *Kepler-438b* orbiting not a red but an orange dwarf star in the constellation of Lyra, which is 470 light years away. It is slightly larger than the Earth and is bathed in 40 per cent more heat from its star than the Earth is from the Sun. Its small dimensions imply that it is a rocky world, and it sits within the Goldilocks zone. Despite being exciting candidates, these worlds will surely not hold the podium finishes for long.

Gliese 832c is not one of the top three potentially habitable exoplanets but this super-Earth is located only 16 light years from the Earth. It is a rocky planet orbiting in the interior of a planetary system thought of as being similar to a miniature Solar System, with a gas giant in the outer reaches. Gliese 832c has a mass 5.4 times that of the Earth and takes 35.68 days to orbit its sun. Given its large mass, it seems likely it would possess a massive atmosphere similar to that of Venus, making it inhospitable for life, although this is not known for sure. Its Goldilocks orbit should allow for liquid water to persist on its surface; the planet's atmosphere, however, would determine whether any such water were usable by life. The number one question is, 'Could a super-Earth such as this support life?' So far, all super-Earths have been found orbiting smaller dwarf stars (yes, this may have something to do with the fact that they are easier to detect), yet to be capable of supporting life, these worlds will need to orbit dangerously closely to their star to maintain the best temperature for liquid water to be stable on their surface. Super-Earths will also have a larger gravity than the Earth. Life inhabiting any oceans would have no problem with this, as the buoyancy of water would balance out the greater gravity of a super-Earth. On land and in the air, however, it would be a different story. We will describe later what kind of life might evolve, should it live on a world with higher gravity and a thicker atmosphere, but the key message is that it *is* possible!

Revealing Red Dwarfs

The vast majority of exoplanets detected so far lie within planetary systems orbiting red dwarf stars, which are smaller, colder and dimmer balls of light than the Sun. For a world to be habitable in this type of star system, it must hug a tight orbit around its sun to keep it warm enough for liquid water to be stable. Such a close orbit means there is a high probability that the planet would become *tidally locked*. This means that one face of the planet would always be looking towards the star, bathed in eternal sunshine, and one would always be turned away, in perpetual darkness, similar to the Moon's experience as it orbits the Earth. This creates a huge temperature dichotomy between the two sides and could produce global gale-force winds. Neither face of these worlds is attractive for life, although there is a sliver of hope. A thin zone could exist, encircling the planet on the boundary between day and night, sitting in the aura of a never-ending twilight that could potentially have a temperature range suitable for life. Sadly, however, intense heating caused by the proximity of planets to their host red dwarfs would nonetheless be a major impediment to life developing in these systems. There would be a very small circumstellar Goldilocks zone owing to low light output from the star, and this light would be shifted into the infrared spectrum, unlike light from our Sun. Finally, if that were not tough enough for life, many red dwarfs earlier in their life are also far more violent and unpredictable than their more stable, larger cousins such as the Sun, in a matter of minutes erupting with flares that double their brightness and send torrents of atmosphere-stripping charged particles towards any unsuspecting nearby planets. You may therefore wonder why we are bothering to look here? First, red dwarf stars are the most common type of star, making up 73 per cent of all those in the Milky Way. Second, they are very long-lived. Red dwarfs could live for trillions of years (in fact, it is thought that no red dwarf has actually died yet) since their nuclear reactions are far slower than those of larger stars, meaning that life would have both longer in which to evolve and to survive. Our nearest star, Proxima Centauri, is a red dwarf like this, and so is TRAPPIST-1 at just 40 light years away. Proxima Centauri has an Earth-sized world sitting at the right distance from its star that it could have liquid water on its surface if it has the right atmosphere. TRAPPIST-1 has seven Earth-sized worlds around it; three of them are worlds that will be vastly different than our own, yet may house life all the same.

Exploring Exomoons

As the name implies, an exomoon is a satellite in orbit around an exoplanet or extrasolar body. The number of natural satellites found in our Solar System orbiting terrestrial worlds and gas giants alike suggests that exomoons should be equally common in other planetary systems. Sizeable moons in a stellar Goldilocks zone could quite possibly outnumber planets. When we focus on moons in our Solar System, there are quite a few with many potentially habitable environments – so why could the same not be true in other solar systems? Furthermore, only four worlds in the Solar System other than the Earth show evidence of current tectonic or volcanic activity, and these objects are not planets but moons. Exomoons are, however, extremely difficult to detect and confirm, owing to their size. Observations from missions such as Kepler have hinted at a number of candidates, including some that may one day turn out to be habitats for extraterrestrial life. As yet, no exomoons have been confirmed, but they are sure to lurk within the vast Kepler data sets, awaiting discovery.

Exomoons may prove to be better candidates for life than exoplanets, as moons can have multiple energy sources. While the habitability of terrestrial planets is generally influenced by the levels of sunlight reaching their surfaces, moons also receive reflected starlight from their parent planet as well as thermal emissions from the planet itself. If a moon were to orbit a planet similar to Jupiter, which is quite possible given how many exoplanets have been classified as hot Jupiters and since Jupiter itself has 67 moons, even more energy sources would be available. A planet such as this would still be shrinking and thereby converting gravitational energy into heat, so it would actually emit more heat than it received from the Sun, providing yet more illumination on to its nearby moons. Besides this, moons orbiting close to a gas giant are flexed by the planet's gravity, providing potential tidal heating as an internal, geological heat source. The distance of the moon to the planet would also play a role in determining habitability, because the closer they lie together the stronger this tidal heating would be. Too close, however, and it is likely that the moon would suffer from a catastrophic runaway greenhouse effect, boiling away surface water and leaving it uninhabitable. These complex interactions between a planet and its satellite could affect the latter's climate enough to make it suitable for life, even if the host planet were completely inhospitable. It seems there is even a Goldilocks zone for exomoons around their exoplanets, as well as exoplanets around their star.

A Pale Blue Dot

If there were an alien civilisation on a planet currently orbiting one of the 100 or so nearest Sun-like stars, what would they see, were they to look towards the Earth? Could they tell it is teeming with carbon-based life forms ready to shake their hand/tentacle? And how is the Earth helping us to identify signs of life in nearby planetary systems? If we look at the Earth from over 4 billion kilometres away, it fits within a single pixel as a tiny pale blue dot, but intelligent alien life could learn a great deal from this infinitesimal speck of light. The Earth varies in brightness over time as clouds and continents move across its surface, so extraterrestrials could infer that this little planetary object has weather, water and even rocky continental bodies. This blue spot on the horizon could also reveal to our inquisitive neighbours what kinds of gases are in its atmosphere. Each gas present will either remove or absorb a separate wavelength of starlight, leaving gaps in the complete rainbow spectrum of light. By looking at the different missing pieces from Earth's spectrum, the aliens could make an educated guess about what comprises Earth's atmosphere. There is plenty of water vapour in our atmosphere, which would suggest liquid oceans are lapping across the planet and therefore that a key ingredient for life, as hopefully they know it, is present. The atmosphere would also be found to contain an unusually high amount of oxygen. Since oxygen is a highly reactive gas it normally combines with other substances and does not exist on its own for very long – it should not really be in Earth's atmosphere at all. But plants and photosynthetic bacteria continually produce oxygen so on Earth, at least, there is always a large amount in the atmosphere. We say that oxygen is a *biosignature gas*: a gas produced by life itself. This would tell the aliens that there is oxygen-producing life, as well as oxygen-consuming life, on the planet.

ET would also be able to see carbon dioxide, methane and other important trace gases in Earth's air. Methane is composed of one carbon and four hydrogen atoms stuck together, and can be a tiny, but powerful, biosignature gas as the product of life as well as one of life's most basic energy sources. Most of our methane comes from countless tiny microbes feasting in the depths of Earth's oceans and swampland as a by-product of their metabolism. It can also, however, be created without any input from life whatsoever. Abiogenic methane arises when volcanically heated water reacts with rocks containing high levels of iron and magnesium. Because of the heating, hydrogen in the water is liberated. This free hydrogen then meets with carbon that has come from carbon dioxide dissolved in the water. The result is methane that has nothing to do with life forms. And yet while not every process that produces methane is generated by life, the overwhelming majority of known sources are alive. Knowing that methane exists in the atmosphere of a planet serves another life-related function as well: it can inform one about the surface temperature. Methane is one of the most notable greenhouse gases. Like carbon dioxide, atmospheric methane acts as a sort of planetary thermal blanket; it wraps around the Earth and absorbs surface radiation that would otherwise make its way into space. In fact, of the two, methane is a far more efficient warming agent and has at least 20–25 times the global warming potential of carbon dioxide. Fortunately for us, however, methane only persists for a few years after it is produced; otherwise things on Earth would be pretty toasty.

Interestingly, and sadly, alien astronomers would also be able to see evidence of humanity itself and the negative effect it has had on the world. The Earth is surrounded by at least 500,000 pieces of greater-than-marble-sized debris or 'space junk' – natural fragments of meteoroids yes, but also leftover remnants of now-defunct spacecraft, satellites and launch vehicle stages. Who is to say whether distant observers could spot this orbiting belt of debris or perhaps even glimpse sunlight reflecting off from it? In the atmosphere, they would see evidence of an industrialised civilisation and its pollution. Chlorofluorocarbons (CFCs), in particular tetrafluoromethane (CF_4) and trichloro-fluoromethane (CCl_3F), are the key ingredients in many mass-produced products on Earth, for everything from holding hair in place to eating holes in our ozone layer. These are compounds that can only be produced by advanced industry. Alien searchers might be tempted to shake whatever passes for their heads in sadness, and move on to contemplate other less-polluted options.

Designer Life

What might life look like on some of these exoplanets, assuming they had sufficient conditions and the time available for complex multicellular plants and animals to arise? If life developed on a world with a lower or higher gravity than that of the Earth, or with no landmass, or it orbited a red dwarf star instead of a Sun such as ours, what adaptations would have developed to enable it to survive? What common traits might we expect to evolve on any planet regardless of the starting conditions? And would we even recognise this life as 'living'?

The Degrees of 'Alienness' ...

There may be a number of ways to arrange and assemble organic molecules to create a living being, but it is incredibly hard for us to imagine one better suited to run and support the functions of life than the system of DNA/RNA storage, the operating manual of a cell and proteins. If any life form arising in the Universe were to be built with a similar sugar-phosphate backbone, then these alien beings would essentially be terrestrial and biochemically similar to us. Many scientists also believe the symbiotic evolution of the eukaryotic cell and the building of multicellular organisms to be all but inevitable.

Owing to the distribution of organic material in the Universe and carbon being the third most abundant element in the cosmos, the first degree of alienness is a population based on the same organic building blocks as Earth-based life, such as amino acids and sugars, but assembled in different ways or using different sets. To recap from Chapter 2, all proteins within life on Earth are built from a combination of just 20 amino acids, yet a catalogue of more than 70 different types of amino acid has been found within rocks from space. Most are alien to terrestrial life. Also, most molecules within a cell have a particular chirality, described earlier. As such there are two possible mirror-image versions of a molecule, which are called enantiomers. All terrestrial life uses one enantiomer – biologically-produced amino acids are left-handed and sugars are right-handed. Perhaps on another world life will use amino acids, but right-handed forms instead of left. And as mentioned in detail back in Chapter 2, life on Earth depends on water and carbon. The most alien life we could imagine would not be based on water and maybe not even on carbon.

We are looking for life in the middle range of alienness, put together using the same building blocks as Earth life but arranged slightly differently. Furthermore, its organic chemistry would have been designed (evolved) to help the organism survive the environment within which it existed. We are not looking nor do we expect to find life identical to that on Earth. For a start, it is a very narrow-minded point of view, but also, if it were found, there would be no way of proving it were alien and not simply terrestrial contamination. Even focusing on the possible variants of life as we know it, there turn out to be a number of familiar features we might not be too surprised to see staring back at us ...

The Predictability of Evolution

The most difficult question about evolution is 'Why?', since this is not posed by evolution itself – it is not working to a game plan or plotted timeline. Instead, the current prevailing conditions represent the only place where evolution occurs, and organisms change in response to fluctuating opportunities or crises imposed by their environment. Evolution is driven by random mutations and natural selection. If an adaptation (or mutation) is useful to a lineage, chances are that it will be preserved and passed on to future generations. Individuals best suited to their particular environment will prevail while those that are not will perish. It is rational to assume, therefore, that an alien organism developing on another world in the distant reaches of our or another galaxy would be subject to the same evolutionary demands as terrestrial biology.

Useful adaptations for life seem to emerge throughout evolution again and again. Evolution is not particularly original but it is innovative, upcycling existing proteins to play new roles and turning previously negative traits into a positive adaptation for survival. This has its limits, however – everything does – and Mother Nature seems to hit upon the same designs over and over again. If it worked well the first time, then why not a second time – Earth's own 'If it ain't broke, why fix it?' philosophy. We call this *convergent evolution* – a process whereby species not closely related live in similar ways and/or in similar environments, and by having to face the same environmental challenges are likely to evolve similar traits. An example of convergent evolution is the similar nature of the wings of insects, birds, pterosaurs and bats. All four designs of the wing serve the same function, to enable the organisms to fly, and are similar in structure, yet each evolved completely independently of the other. Convergent evolution is extremely common on Earth. In their own exploration of what life might lie waiting in the Universe, popular British science writers Jack Cohen and Ian Stewart, a reproductive biologist and mathematician respectively, listed the *four F's* of universal evolution in terrestrial biology: fur, flight, photosynthesis and ... sexual reproduction.

The late evolutionary biologist Stephen Jay Gould, in his 1989 book *Wonderful Life: The Burgess Shale and the Nature of History*, discussed the Burgess Shale fossils described back in Chapter 4. You may recall that they are a collection of strange alien-style life forms that inhabited the Earth's oceans about 520 million years ago. Many species from this time in the Cambrian have since died out because they were not fit enough to compete for survival or were in the wrong place at the wrong time during volcanic eruptions, asteroid impacts or other extinction events. Gould theorised that life today would have been very different had history unfurled in another way, that life is a result of the outcomes of past accidents – *historical contingencies*. Random mutations and chance extinctions would build on each other, driving the evolution of life down one path or another. In Gould's view, the existence of every animal, including humans, was a rare event that would have been unlikely to recur if the tape of life were rewound to the Cambrian Period and replayed. For example, it is widely believed that the chance asteroid impact 65 million years ago that killed off the dinosaurs allowed mammals to arise and humans to become the dominant species on the Earth. Without this impact, would we even be here?

Life is a lottery of convergence and contingency, a potluck of inevitable adaptations and some lucky chances. The same interactions between convergence and contingency may play out on other planets, with many features of an alien being partially predictable, while others can only be based upon quirks in their evolutionary and environmental history. If extraterrestrial life has faced similar evolutionary pressures as life on Earth, future humans may discover aliens that have convergently evolved to resemble us, and have intelligence similar to ours. On the other hand, if contingent events build on one another and are responsible for driving the development of life down unique paths, as Stephen Jay Gould suggested, then extraterrestrial life may be remarkably strange. This means, however, that in order for us to speculate on what alien life might look like we should pay attention to the convergent adaptations life has created, as the number of lucky breaks is probably infinite. Furthermore, if an alien were evolving in an environment very different to the Earth, then it might have developed a number of unique design solutions to allow for its survival. One of the most important questions in evolutionary biology is what features of organisms are universal and might be expected to re-appear every time life arises or is restarted, regardless of the planet or moon it finds itself on.

A Body and Brain

In popular culture, we predominantly portray ET as humanoid with four limbs, standing upright and with a forward-facing head. Although this probably is not alien enough, we know it is as a result of the budgets placed upon costume departments and that it is more engaging and believable to witness Captain Kirk talking to a human-like alien than a gelatinous blob of goo. However, there is some truth behind the idea. On Earth, birds, reptiles, fish and insects are all constructed with bilateral symmetry. This means the left and right halves of their bodies are reflections of each other. Alien life may quite possibly have chosen the same layout. Yet it could also have a different fundamental plan, such as the radial symmetry of jellyfish, whereby it has no left or right side, only a top and a bottom. Most vertebrates have an entrance and an exit – mostly separate, sometimes shared – and some form of *skin* to contain all their organs. There is also a common need for lungs or gills in creatures, once they reach a certain size, in order to be able to access oxygen. And before you ask, there is good reason to suppose that alien life would enjoy a deep breath of oxygen as much as we do. Burning organic carbon-based fuels, such as glucose in oxygen, provides greater levels of energy – enough to satisfy the enormous power demands of animal life.

Larger animals colonising the land would also require some kind of support or scaffolding to hold their body together against the force of gravity, as well as a frame against which muscles can push. They will therefore most likely have a skeleton that can either be an 'innie' like ours or an 'outtie' like that of crustaceans. But what about limbs? These are obviously needed in some form for grasping objects and in most animals for movement, but how many would an alien have? Humans and all land vertebrates have a four-limbed body plan (two arms and two legs), as a result of contingent evolution – our fishy ancestors by chance had two pairs of lobed fins, and so we have four limbs. An alien ancestor could just as easily have had three pairs of limbs, creating hexapod descendants along the lines of terrestrial insects, and maybe even have adapted these limbs so that the front pair are no longer used as legs but as arms or claws similar to those of crabs on Earth. Six legs would be a very fortunate adaptation for life on a world with stronger gravity, supporting the movement of much heavier, although not necessarily larger, life forms. At the end of our limbs, humans have 10 fingers and 10 toes, but we could balance or grasp objects just as well with four or six digits, just as long as we kept our opposable thumbs. Who knows how many fingers an alien might have – the point is that the chances are they will have them.

The Face of Life

The development of the eye is seen as a universal evolutionary feature. It makes good design sense to have a way of *seeing* where you are going and, to that end, placing the eyes at the front of a head. It is clever to put the best surveying organ available close to the brain and for it to be protected by a hard shell (the head). The human eye is an exquisitely complicated organ that acts in a similar way to a camera, collecting and focusing light and converting it into an electrical signal that the brain translates into images. Instead of photographic film, it has a highly specialised retina that detects light and processes the signals using dozens of different types of neuron. Humans and most other vertebrates, and cephalopods (which includes octopuses, cuttlefish and squid) have *camera-type* eyes, which took shape in fewer than 100 million years. Eyes first evolved during the Cambrian Explosion from light-sensitive proteins based in a single illuminated spot, and were used to monitor circadian (daily) and seasonal rhythms. They then evolved into light-sensitive pits, into compound insect eyes and finally our optically and neurologically sophisticated eyeballs by 500 million years ago.

Complex, image-forming eyes have evolved independently some 50 to 100 times, which is not particularly surprising, given that any model or method allowing organisms to see and rapidly focus on an image would confer an enormous evolutionary advantage for survival, regardless of whether they are based on land, air or sea. Many genes are responsible for making the eye. One holds instructions for making light-sensitive pigments, another provides information for making the lens, and other genes orchestrate it all, directing various parts when and to where they need to be assembled. These are called *master control genes*, and for eyes the most important one is *Pax-6*. The ancestral Pax-6 gene controlled the formation of the first very simple eye and still controls today's most complex incarnations. Given that a variety of eye-type structures has arisen independently numerous times, it seems highly likely that an alien life form will have eyes. To be at the pinnacle of evolutionary adaptation, it would quite possibly have an eye similar to that of an octopus – one step better than the eye of a human as it lacks the *blind spot* created where the optic nerve leaves the eyeball.

Sensory organs give animals an acute awareness of changes in the environment around them and throughout their bodies, so that they can physically respond. A very satisfactory evolutionary adaptation, these touchy-feely organs enable animals to avoid hostile environments, sense the presence of predators and hunt down sources of food. Animals can perceive a wide range of stimuli that includes touch, pressure, pain, temperature, chemicals, light, sound, movement and position of the body. Some animals can also sense electric and magnetic fields. The *special* senses of smell, taste, sight, hearing and balance are particularly useful, and the organs used for these relatively complex. Life forms need to process this sensory information as fast as possible so most have a centralised nervous system, and to reduce data transfer times nature houses this near to the organs. It seems, therefore, that the development of a head at the front of the body may prove to be universally essential among higher animals, although the positioning not so much.

Skin and Bones

Human skin colour ranges from the darkest brown to the lightest pinkish-white, even yellowish, hues. Human skin pigmentation is the result of natural selection; it evolved primarily to regulate the amount of ultraviolet radiation penetrating the skin, controlling its biochemical effects. There is a direct correlation between the geographic distribution of UV radiation (UVR) and the distribution of indigenous melanin skin pigmentation around the world. Areas that receive higher amounts of UVR, generally located closer to the Equator, tend to have darker-skinned populations. Areas far from the Tropics and closer to the poles have a lower intensity of UVR, which is reflected in lighter-skinned populations.

Alien life would not, however, necessarily follow the same human melanin spectrum. It might more closely resemble that of terrestrial plants than Earth's vertebrates. Might aliens live up to their fictional reputation and be a greyish green? This colouration of particular life forms on Earth is a consequence of two things: in plants it comes from the photosynthetic pigment chlorophyll; and in animals from the need to be camouflaged and remain hidden within our planet's vast vegetation. Some biologists believe that chlorophyll may be a universal molecule, used by any number of organisms to soak up the light of their parent star. However, the wavelength of light emitted by a star varies, depending upon its temperature. A star cooler than the Sun, such as a red dwarf, will shine more in the infrared than in the visible portion of the electromagnetic spectrum, so any photosynthetic pigment would need to be tuned to absorb this differing wavelength of light. It might well be perfectly camouflaged to its own vegetation but to our eyes would not look green. If an alien did appear green to us, then it is a pretty fair bet that its sun would be very similar to ours. An alien might also appear green if it used chlorophyll to provide the organism with a source of nutrients. Unfortunately, although perfectly adequate for the energy budgets of plants, photosynthesis cannot provide nearly enough energy to meet an animal's high demands, especially powering muscles and a brain. Complex life must be carnivorous, devouring the nutrients contained within plants and other animals (which themselves may feed on photosynthetic plants) to draw the energy it urgently needs. There is nothing to say, however, that a carnivorous life form could not have a chlorophyll-rich skin for times of famine or for use as a top-up source of energy and nutrients for survival in harsh or seasonally challenging environments.

An internal circulatory system is needed in any life form to move nutrients effectively to where they are needed, as well as to remove waste products. For terrestrial life, rich red blood is the main mechanism, although in actual fact blood can be a number of colours, including colourless, as a result of the specific chemicals it carries. Humans and most other vertebrates use the iron-containing protein haemoglobin, and so have red blood. Haemoglobin is a respiratory pigment and plays a vital role in the body, ferrying oxygen to cells and helping waste carbon dioxide return to the lungs where it can be exhaled. When it is oxygenated and full of freshly breathed-in oxygen, it is bright red in contrast to when it is deoxygenated, when it is a deep, dark red. It is a commonly held myth that deoxygenated blood is blue – after all, if you look through your skin at any of your veins carrying deoxygenated blood away from your body's cells, they have a definite blue-grey hue. However, this appearance is in fact caused by the interaction of light with both the blood and the skin and tissue covering the veins. There are some creatures, however, for whom blue blood is the norm. Unlike haemoglobin, which is bound to red blood cells, *haemocyanin* can float freely in blood and contains copper instead of iron. When deoxygenated it is colourless, but blue when carrying oxygen. This is the shade of blood found in spiders, crustaceans and some molluscs, octopuses and squid. Green blood also exists in some segmented worms, leeches and sea cucumbers, which contain *chlorocruorin* or a mixture of haemoglobin and chlorocruorin. Finally, violet blood has been found within marine worms and brachiopods as it contains *haemerythrin* as the oxygen transporter. What is perhaps most interesting about the varying colours of blood is that it showcases evolution coming up with different solutions to the same problem – in this case, how to transport oxygen.

One adaptation I consider particularly useful, and could easily imagine any alien life form adopting, is being streamlined. It turns out that there are not many ways of remaining streamlined while pushing yourself through the water. On Earth, salmon, whales, penguins and water boatmen all come from very different lineages and yet have independently converged on the same body plan of a sleek bullet-like shape helped by fins and flippers. There is no reason why this may not have occurred elsewhere as well. Perhaps on another world, jet propulsion would be a more dominant mode of movement. Terrestrial squid contract an outer cavity to move backwards in pulses and it has been speculated that animals larger than sharks on another world could also propel themselves this way by contracting water or air through a hollow tube running the inside length of their bodies. The need for an aerodynamic body is not just good for swimming, but for the one thing we all wish we could do – fly.

A Flight of Fancy

The evolution of flight might be expected or even inevitable on any terrestrial planet. But before there was to be flight, life needed wings. The way in which these evolved on the Earth is not precisely known although some theories perceive an evolutionary step from arms used by bipedal animals to leap into the air to capture small prey, which evolved into large wings to assist in said leaping. Perhaps wings arose from gliding ancestors who began to flap their gliding structures in order to produce thrust so as to move faster and further. We know that flight evolved millions of years ago in all of the groups that are capable of flight today and the reasons differ depending on the species in question. These range from helping them to escape from predators or catch flying or speedy prey, aiding movement from place to place (leaping or gliding), freeing the hind legs for use as weapons, or gaining access to new food sources or an unoccupied niche.

Gliding – a controlled descent using gravity as the driving force – has evolved many times and in many different groups including frogs, lizards, snakes and several different mammals, and even the seeds of some plants. Gliding is also known among some fossil reptiles. On the other hand, active flight, during which flapping powers an organism through the air, has only evolved four times in nature: once in arthropods and three times in vertebrates. It has only become extinct once … in the *pterosaurs*. The first animals to evolve flight were insects some 410 million years ago, and the pterosaurs were the first vertebrates to evolve active flapping flight, although the origin of this adaptation is something of a mystery. The transformation from non-flying, perhaps gliding, animal to fully-flying pterosaur probably

EXTRATERRESTRIAL WORLDS 215

occurred in the forests of the Middle Triassic. Unfortunately, this environment rarely yields fossils, so the search for the oldest or even original pterosaur may be in vain. Shortly after the Cretaceous, bats appeared – the first active flying mammals – and about 50 million years after that another group of vertebrates achieved the ability to fly. Instead of evolving wings directly, this group used its ingenious brain to skip that step and build machines to fly for them: this was our species.

On Earth, the energetics of staying airborne impose stringent limits on the size of flying animals, but this may not be the case on planets and moons with different gravities and thicknesses of atmosphere. Planets smaller than the Earth have a lower gravity and so tend only to hold on to a thin atmosphere. This would mean that flapping wings would generate less lift and flight would be rather difficult, even though the physical pull of gravity holding the organism to the ground would be reduced. Wings with very large surface areas and spans of up to 75cm (2.5ft), similar to those of prehistoric dragonflies, might help. Flight would actually be easier on a larger world with a stronger gravity, as its pull on the atmosphere would be stronger, and the air would be denser, maintaining a thicker, more flight-friendly environment. Animals as massive as elephants would be able to glide through the air and *sky-whales* could dominate the clouds, carried on thermals and slowly flapping their immense wingspans.

If I had to design a flying alien species, I would give them wings modelled on those of a bat. Bat-style wings have evolved several times in mammals. They use up less energy during flight as a result of flexible skin membranes and are held together by more than 24 joints. As a matter of fact, bats are operating with the same skeletal structure as humans. Every joint in the human hand is found in a bat's wing, and indeed a couple more. The wings provide more lift and less drag, thereby increasing manoeuvrability.

Lighting up Life

Bioluminescence is the production and emission of light by a living organism. It is a form of *chemiluminescence,* whereby light energy (luminescence) is released as the result of a chemical reaction. Bioluminescence occurs widely among animals, particularly in the open sea, including jellyfish, comb jellies, crustaceans and cephalopod molluscs. This characteristic is also present in microorganisms including some bacteria and fungi, and terrestrial invertebrates such as insects. The most famous, perhaps, are the dinoflagellates frosting breaking waves in a blanket of phosphorescence and weaving light ribbons across the sea, while on land fireflies are amazing to watch as they dance around on a warm summer's evening. Fireflies actually provide so much light that they were once carried underground in jars to provide light for miners working deep in the bowels of the Earth.

Bioluminescence can serve several functions depending upon the needs of the organism using it, which can range from *counter-illumination camouflage*, in which the animal matches the overhead environmental light as seen from below (such as in the oceans), to attraction of a mate or, conversely, attraction of prey, snaring food in glowing threads and webs. Furthermore, it can be used in defence and in warnings, for communication, mimicry and, of course, simple illumination.

In some animals, the light is not their own, coming instead from symbiotic organisms such as *Vibrio* bacteria. Other organisms on Earth contain a light-emitting pigment called *luciferin* and the enzyme *luciferase*, which reacts with oxygen to create light both within and outside cells. In evolution, luciferins generally vary very little, with one in particular, *coelenterazine*, present in the light-emitting pigment of nine ancient groups of organisms. Not all manufacture coelenterazine themselves – some obtain it through their diet. Overall, bioluminescence has arisen over 40 times in evolutionary history and is found within at least 70 genera of squid. Most marine light emission is in the blue and green light spectrum. However, some loose-jawed fish emit red and infrared light and the genus *Tomopteris* emits yellow light.

An Alien Greenhouse

Unsurprisingly, convergent evolution is also common in the plant world and can give a number of clues as to what vegetation might arise on an exoplanet or moon. Plants, especially those on land, need to satisfy four fundamental constraints to survive. They must be able to catch as much light as possible from their Sun to allow for photosynthesis to work and they must be able to disperse pollen or seeds as far as possible to ensure the survival of their species. They also need to ensure mechanical stability so as not to topple over and must have ways of retaining, or not losing too much, water. Depending upon the demands of the local alien environment, plants might meet these requirements in a number of ways. Slow-growing vegetation arising on a world with a low availability of light would be likely to develop wide, flat canopies to maximise the amount of light intercepted (although this could cause a stability problem, increasing the risk of being uprooted in strong winds); while those on water-poor worlds might evolve in a similar way to squat cacti with highly modified leaves, such as spines. As well as defending against hungry herbivores, spines help prevent water loss by reducing airflow close to the cactus and providing some shade. As such, the shapes of trees and plants in an alien forest are likely to resemble those found in similar environments on the Earth and might be recognisable to us. On a planet with high wind speeds, mechanical stability would be of paramount importance and trees may resemble terrestrial firs, or seaweed with its submissively flexible stem. Worlds with a higher gravity might have low-lying trees with stout trunks and fewer branches. Potentially on these worlds, plants might evolve novel ways to reach the sunlight, reproduce and access water. Perhaps instead of desperately trying to grow towards the sky, alien plants would simply float upwards. Photosynthetic plants on Earth use the hydrogen produced by the splitting of water molecules to generate food, and release oxygen as a waste product. If an alien plant released this hydrogen inside an inflatable sac, it might float into the sky like an airship to find what it needed, anchoring itself to the ground with a vine. Going one step further, this vine might have the ability to detach during reproduction to allow the plant to be carried in the wind and disperse its seeds across vast distances. This adaptation is found in the seas on Earth when kelp forests release seeds contained in small flotation bladders, pumped full of oxygen or carbon dioxide.

We would hope to recognise the shapes and even life strategies of a plant or tree on another world, but would we recognise its colour? Would an alien biosphere be as green as the vast majority of healthy vegetated areas on Earth? If alien plants were found on a world orbiting a Sun-like star and use chlorophyll as a pigment, then quite possibly yes. However, even the Earth hosts a variety of organisms other than green plants that photosynthesise other than green plants. On land, plants may also display foliage that is red, yellow, orange, cream, purple or demonstrate a variety of other effects, while underwater algae and photosynthetic bacteria are also found in a wide palette of shades. The dominant colouration of extraterrestrial foliage will depend on how alien photosynthesis evolves in response to the spectrum of light received from its parent star, combined with the filtering effects of the planet's or moon's atmosphere (which may be very different to our own) and, for aquatic creatures, of liquid water (or the liquid they are residing in). In general, plants on Earth use the broad spectrum of visible light (red–orange–yellow–green–blue–indigo–violet) from the Sun, profiting most from the blue-green range. The Sun transmits predominantly red *photons* (particles of energy from light or other electromagnetic radiation), useful for their quantity, though lower in quality than blue photons, which supply more energy. Green photons in between are lacking both in such energy and numbers, so vegetation on Earth has evolved to screen much of these and thus reflect green.

As always, the answer to the puzzle of what vegetation may be found on other worlds lies once more in the stars. Astronomers grade stars according to their colour, which is related to their temperature, size and longevity. It is clear that only certain types of star are long-lived enough to allow the evolution of complex life to occur. From hottest to coldest these are the F-, G-, K- and M-class stars (our sun falls into the G category). F stars are larger, burn more brightly and bluer, and exhaust their fuel over a couple of billion years. K and M stars are smaller, dimmer, more red in colour and survive longer. We know that light of any colour from deep violet through to near-infrared could power photosynthesis. For another Earth-type planet orbiting around a G-class star, we can comfortably predict green, yellow or orange plants. Around stars that are hotter and bluer than our Sun, the abundance of blue photons would be so overwhelming that plants might need to shield themselves against it, using a pigment similar to anthocyanin, so that they would reflect blue. Since, however, they would still try to absorb a percentage of the energetic blue light, they would actually appear green to yellow to red. The span of M-star temperatures would make possible a wide spectrum of colours in alien plant life. A planet around a cooler M star, such as a red dwarf, would receive about half of the energy that Earth gets from the Sun, with less visible light and a spectrum closer to the near-infrared range. With less energy available, plants might try to absorb as much as possible, and with little light left over to reflect, appear black.

Once we have observed and hopefully recognised life on another world, be it animal, plant, microbe or maybe even mineral, what do we do next? Why ... we try and say hello, of course.

CHAPTER NINE

Is the Truth Really Out There?

At a 2015 press conference, NASA's Ellen Stofan and John Grunsfeld predicted that astrobiologists would finally detect alien life within the next 20 years. Once, this would have been viewed as a rash statement, but today, weighty evidence suggests that warm, wet environments habitable by forms of life we would recognise exist throughout the Universe. Our understanding of life's fingerprint and evolutionary manoeuvres grows by the day. The potential existence of sentient aliens and extraterrestrial cultures, on the other hand, remains controversial. It is the loneliest question in the cosmos – are we the only life forms in the Universe? The enormous size of our Galaxy, containing potentially 400 billion stars, makes it difficult to imagine that our planet is isolated in its status as the host of intelligent life. Somewhere out there, surely, are other life forms with whom we might be able to communicate – at least at some level. If they are out there, why have they not contacted us and are we even capable of detecting their presence? One assumption is clear – that any civilisation, by the mere fact that it has become a developed society, is intelligent. So what exactly is the fingerprint of intelligence? Who decides what intellect actually is? And do we as a species have enough of it to be able to find and recognise it elsewhere?

The Search for Answers

Our interest in whether or not life is out there stems from our ability to see the stars, and recognise that we are but one lowly planet orbiting just one of them. The notion that vast galactic empires may be conducting their own business in outer space remains conjecture that is based on apparently logical conclusions drawn from our own origins. We know that microbial life evolved on the Earth at least, and that life then took many twists and turns via natural selection and some chance events to develop complex multicellular bodies and brains, societies and finally technologies that could provide a means of transport to other planets – maybe one day to other stars. Who is to say this might not equally occur on any of the billions of other habitable planets we believe twirl out in the cosmos? The search for sentient life is not just scientific but also spiritual, based on a belief it is out there somewhere.

We want to answer one of the world's oldest questions: 'Are we alone in the Universe?' There was intensive debate about this issue even in antiquity. The atomists, who, as their name suggests, perceived correctly that atoms were the basic material of which everything is made up, believed in a plurality or a number of worlds. Yet, the Aristotelians held the opinion that we live in a closed cosmos with Earth at its centre. In the medieval age, philosophy and theology were dominated by the Aristotelian world view, until finally the Copernican revolution paved the way for the belief, and later proof, that there are many solar systems similar to ours in the vastness of space. Two widely assumed principles supported the idea that life is abundant in the Universe: the principles of *plenitude* and *mediocrity*. The former holds that a Universe made by a perfect Creator should be as rich as possible – and what universe could be richer than one that gives birth to and provides a home for a plenitude of life? The latter suggests that every place in the Universe that has similar molecules and shares similar laws of physics to Earth, would be likely to develop along a more or less similar route as our planet. The resulting multiplicity of Earth-like planets is a principle held by many in modern science. Nevertheless, whether or not that principle also applies to the existence of life remains an open question.

The Oddity of ET

In an attempt to quantify the possible number of cultures present in our Galaxy, the astronomer Frank Drake formulated a completely unsolvable equation in 1961 that, despite its unknowable and currently improvable answer, has gone down in history and is taught on all planetary science courses. The equation below is not some mind-boggling mathematical calculation – just a useful, relatively simple tool to help us contemplate the variables we must incorporate when considering the question of life elsewhere, and it is actually quite intriguing. Here it is:

$$N = R^\star \times f_p \times n_e \times f_l \times f_i \times f_c \times L$$

We are trying to figure out the value of **N**, which represents the number of civilisations in our Galaxy with whom radio communication might be possible. To arrive at this number, a variety of factors must be taken into account. Starting at the beginning, $\mathbf{R}\star$ is the average rate of star formation in our Galaxy. Estimates for the number of stars in the Milky Way vary from a low of 100 billion to a high of 400 billion. Estimates for the age of the Milky Way also vary from an infant of 800 million years to a grandfather of 13 billion years. If we go with the lowest star count and the oldest age for the Galaxy, the average rate of star formation works out at 7.7 new stars per year. If we go with the highest star count and the youngest age for the Galaxy, the average rate of star formation becomes 500 new stars per year. The rate of star formation in the galaxy is not constant over time, however; stars were formed at a much faster rate in the Galaxy's earliest moments, and not all stars are created equal or deemed useful for hosting life. Today, estimates for the overall rate of star formation range from 5 to 20 new stars per year and the rate of formation of Sun-sized stars, around which intelligent life may arise, is in the order of one per year.

The next term, $\mathbf{f}_p$, is the fraction of those stars that have planets. At the time Drake wrote his equation, this was a complete unknown. Since then, nearly 2,000 exoplanets have been found orbiting a variety of stars. Estimating the total number of planets in the Universe is difficult, but it is possible that, in the Milky Way, each star has an average of 1.6 planets – yielding 160 billion alien planets in our home Galaxy alone. This number is making the chances for finding communicable life look a little better. The third term, $\mathbf{n}_e$, is the average number of planets that can potentially support life per star that has planets. In his original equation, Drake optimistically assigned a value of 2 to this parameter, meaning that he proposed on average two Earth-like planets per star for those stars with planets. The answer to this remains unknown, but out of the thousands of exoplanets discovered, small, rocky worlds similar to our own are popping up everywhere, and some of them may be capable of hosting life as we know it. The last four parameters, f_l, f_i, f_c and L, are, as you may have guessed, also not known – perhaps forever unknowable – and are very hard to estimate. They are in order: the fraction of planets that actually go on to develop life at some point, the fraction of those that actually go on to develop intelligent life, and the fraction of those intelligent civilisations that develop technology that releases detectable signs of their existence into space; finally, L is the length of time for which civilisations exist.

The most contentious is f_i, unsurprisingly. We have only one example, Earth, where life is abundant and humanity has reached a level of technology that allows it to scour the Universe for other life, and can broadcast its own existence into space. Is human-like, technological intelligence likely to be common across the Universe? Are we merely an evolutionary blip? Or is intelligence something that the entropy-driven, complexity-producing Universe will inevitably converge on? It may even be that Earth (and all intelligent life on it) is an early bloomer and that we are not hearing from advanced alien civilisations because the Universe has not had the time to spawn other habitable worlds on which they could flourish.

In case you are wondering, the estimated solutions to the Drake Equation range from 0 intelligent, communicating civilisations in the Galaxy, to 10,000 – not especially illuminating. We may be one of a myriad of intelligent species or we may be alone. Even if one day we were to receive signals from another intelligent species, the problem would remain of how to reply and stay in contact over such measureless distances, even ignoring the obvious barriers of communication media and language. It could well be that extraterrestrial life exists in parts of the vastness of the Universe that are beyond the possibility of contact or even of mere observation. Yet finding life outside the Earth – even non-intelligent microbial life – would be an important step towards a better understanding of the Drake Equation, and take us further along the yellow brick road to communication.

Conversations with my Cat

As far as we know, no cat can compose an email, no whale can sing to us in our own language and no bird can solve mathematical equations. Only humans can perform such intellectual feats, presumably, some may say, because we are smarter than all other animal species – at least by our own definition of intelligence. Just as with life itself, there is no globally accepted definition of intelligence. It has been described in many ways, from a capacity for logic, to abstract thought, understanding, self-awareness, communication, learning, emotional knowledge, memory, planning, creativity and problem-solving. At its most simple, however, intelligence is the ability to identify and receive information and retain it as knowledge for later use: now the scope for recognising other intelligent species on Earth widens.

Life based on one basic blueprint has existed on Earth for 3.7 billion years – more than a quarter of the age of the Universe – and it took 1.8 billion of those years for the first multicellular versions to appear. Life on Earth is divided into bacteria and fungi, plants, and animals, but nervous systems and brains only developed in animals. Of these animals, some 60–80 lines developed, but only in the chordate vertebrates did intelligence appear. Within the vertebrates, such as fish, amphibians, reptiles, birds and mammals, higher intelligence only developed in mammals. The hominids first appeared some 3–6 million years ago, but *Homo sapiens* evolved only in the last 200,000 years, producing a number of different civilisations, with only the most recent developing technology capable of seeking out other intelligent life. This last development of space-voyaging folk took place only in the last 100 years of the immense 4.55-billion-year lifespan of the Earth. The evolution of hominid intelligence is attributed to specific environmental challenges and it is a misinterpretation of evolutionary theory to see this rise of intelligence in us as a *necessary* process. Should we accept that it was pure chance and a lucky mutation or two? It could have appeared in fish or dinosaurs, or may not have arisen at all, but intelligence is the only adaptation to have allowed a single species to establish complete domination over the rest of the natural world. Will humanity with all its knowledge survive 100,000 years from now? We as a species are good at reproducing to ensure the longevity of our race, but are yet to prove we can use our gift of intelligence to survive across the ages.

Cosmic Conundrums and Fermi's Paradox

This hope of the existence of long-lived alien civilisations capable of first contact is what drove the Italian physicist Enrico Fermi to utter his famous phrase, 'Where is everybody?' In 1950, amid a spate of rumours of flying saucers crashing into New Mexico, Fermi reasoned that if only one in a million stars in the Milky Way had planets with intelligent beings and they began space travel, then within a few tens of million years they should have spread throughout all regions of our Galaxy, including our Solar System. So why, he wondered aloud, have we not seen them? Why have they not made their presence known? This became known as the *Fermi Paradox*. If Earth is not unique in having intelligent life, then civilisations should already have evolved many times over in the Galaxy, since there are billions of stars older than the Sun. If any of these civilisations wished to colonise the Galaxy, they could have done so by now. There are two answers to his question: (1) we are alone, or (2) we are not alone, and the first solves the paradox. Fermi was obviously posing the question because of his belief that we are not the only intelligent life forms floating in the sea of space. A number of solutions has been proposed, none of them provable yet, of course, but an answer I particularly like is that we are akin to animals in a zoo. Humans are bound by our current level of technological intelligence to travel only within our own Solar System. Any civilisation able to overcome these limitations would need to be vastly more advanced – so advanced, in fact, that they may see striking up a conversation with us with our current science, technology and social abilities, as futile, as we would view trying to chat to an ant colony scurrying around an anthill. Hence, we have heard no news. Perhaps more advanced civilisations are waiting until we are worthy. Or, if they don't exterminate us for being so inferior, they may assign us (or already have assigned us) to conservation and study in a galactic zoo with the aliens as our keepers. Perhaps they are reading this and laughing at us for not already realising.

Barriers to Life

The Great Filter is a probability barrier, a concept that some mechanism we cannot yet understand may have prevented life, or will prevent life in the future, from expanding into the Universe. We think there may be one or more highly unlikely evolutionary stages whose occurrence is paramount for an Earth-like planet to form and to produce an intelligent civilisation of a type that would be visible to us with our current observation technology. If you begin with trillions of potential germination points for life, and end with a total of zero extraterrestrial civilisations that are observable – *something* is happening *somewhere* along the line to affect (arrest or slow down) evolutionary development. The critical evolutionary step(s) towards intelligent life must be essential enough, yet unlikely enough, that even with many billion rolls of the dice one ends up with nothing: no aliens, no spacecraft, no signals. This very powerful Great Filter can change the course of history. So what could this Great Filter be? And importantly when did it happen? There are three possibilities: it already transpired in the distant geological past; it has not yet taken place; or ... it is happening right now.

If the filter were in the past, this would explain why we have not yet come across aliens, because if the rise of intelligent life on any one planet is sufficiently improbable, then it follows that we are most likely to be the only such civilisation in our Galaxy or even in the entire known Universe. A potentially unique and extremely improbable step occurred at some point in our history that allowed for intelligent technological life to arise on the Earth. If the Earth movie were stopped and played again from the start, the rise of sophisticated intelligence might not happen in its re-run. Evolutionary biology, at least for the moment, does not enable us to calculate the probability of the evolution of intelligent life on Earth, but we know that some events were critical to its success and may therefore be good candidates for a Great Filter that led down the path to intelligence. One criterion is that the Great Filter must only have happened once (as there is only one example of intelligent life), since features that have evolved multiple times on Earth are demonstrably likely to occur on other worlds. The evolution of flight, sight and limbs, which have all occurred on Earth several times, are ruled out as possible candidate events. Another possible Great Filter is that an event took a very long time to come to pass even after the perfect conditions for it to occur were present. The emergence of life in the first place is an example. As far as we know, the transformation from building blocks into a reproducing, metabolising organism may have occurred only once and have taken hundreds of millions of years to get going, even after the planet had cooled sufficiently to enable a wide range of organic molecules to be stable. It actually took 1.8 billion years for prokaryotes to evolve into eukaryotes, which is quite a long time. This transition is a good candidate for being the Great Filter.

A worrying possibility is that the Great Filter lies in our future. This would mean that some cataclysmic unknown event prevents almost all technological civilisations at our current stage of development from progressing to the point at which they engage in colonisation of space and make their presence known to other technological civilisations. For example, it could be that any sufficiently technologically advanced culture always discovers or creates very powerful weapons and causes its own annihilation. We will never know whether this is the case or not until it happens to us, or until we chance upon another civilisation at this point in its own evolution. A final thought to ponder is whether the Great Filter is happening right now! The greatest obstacle to exploration of the cosmos is the distances involved and the ability of humans to undertake and survive the journey. Could distance and physical frailty in fact prove a Great Filter, and indeed prevent other species from contacting us?

The Search Begins ...

The vast immeasurable expanse of the Universe and the enormous impact that first contact would have on our world has led a few brave souls to take on the scientific challenge of searching the cosmos for intelligent beings. What drives them is the belief that humanity is a normal outcome of physics, chemistry and biology. Since the 1950s, a small number of astronomers across the world have risked ridicule by joining in a search for other sentient beings. In 1984, they founded the *SETI Institute* (the *Search for ExtraTerrestrial Intelligence*), based in California. Utilising massive arrays on Earth, as well as space telescopes, SETI scans the heavens for any indication of radio or other communications beamed in our direction. At the time of writing, no verifiable signal has yet been received. The Universe may be teeming with life, but it is currently preserving its secrets.

Project Ozma

Between April and July 1960, Frank Drake, who wrote the quirky unsolvable equation presented earlier, undertook humanity's first effort to identify radio transmissions from alien sources. Based at the *National Radio Astronomy Observatory* (*NRAO*) in Green Bank, West Virginia, *Project Ozma* was born, named after the queen of L. Frank Baum's fictional land of Oz, described as 'very far away, difficult to reach, and populated by strange and exotic beings'. Drake initially directed his search towards two stars, similar in age to our Sun, known as Tau Ceti and Epsilon Eridani, located some 11 light years (66 trillion miles) away. For six hours every day, Project Ozma's radio telescope scanned frequencies emanating from regions of cold hydrogen gas, looking for repeat sequences of pulses or series of prime numbers such as 1, 2, 3, 5 or 7, that might indicate some form of artificial intelligent message. A thrilling red herring early in the programme proved to come from a now not-so-secret military experiment; besides that, there came only static. Nonetheless, the pioneering Project Ozma generated enormous public interest, and made the search for the technological fingerprints of civilisations on other worlds scientifically feasible.

Prime Numbers and Laser Pulses

In the story *Contact*, published in 1985 by Carl Sagan, a young SETI researcher finds strong evidence of ET life within a radio transmission and is chosen to represent humanity to make first contact. The signal is a repeating sequence of prime numbers apparently sent from the star Vega, and contains 60,000 pages of data outlining plans to build a machine to allow one human occupant to communicate with the alien life forms. This story, written by an astrobiologist and SETI researcher himself, was based on real techniques, real theories and even real scientists.

Jill Tarter (who gave a fantastic TED talk on this topic) acted as Sagan's muse for the novel and was portrayed in the film adaptation by Jodie Foster. The book draws on the idea that alien signals could take many forms – from radio and light signals to laser pulses and even genetic manipulation. We can still only guess at the means by which a more advanced civilisation might choose to make contact.

So far the hunt for alien signals has mostly used radio waves, based on the theory that radio is a relatively easy and cheap way to send signals long distances through space. The SETI Institute uses powerful radio telescopes on Earth to search for signals focused at a single spot on the radio dial, called *narrow-band signals*, and those that repeat in a mathematical pattern. Huge numbers of natural bodies make radio noise, such as small pulsars (pulsar actually being short for 'pulsating radio star'), although not in regular arrangements; yet the only thing that makes a narrow-band signal, as far as scientists know, is an artificial transmitter.

The SETI Institute's *Allen Telescope Array* (*ATA*) is solely dedicated to the search for signals broadcast by intelligent alien life. Located at the Hat Creek Observatory in the Cascade Mountains of California, the ATA is an ambitious array of multiple small dishes whose power can be combined to form the equivalent of a single large-dish antenna, but positioned in several different directions to make a more powerful radio ear than ever before. Because of its ability to study many areas of the sky at once, every day of the week, it has the ability to listen to a truly significant sample of the cosmos.

Exquisitely sensitive, the ATA could detect emissions from powerful radars equivalent to those we have on Earth at a distance of dozens of light years. Any society even slightly more advanced than our own could manage a deliberate radio transmission that the Array could tune into. For SETI researchers, it is only a matter of aiming the antennae in the right direction and establishing the correct frequency. Among its targets are the exoplanet candidates discovered by NASA's Kepler space telescope. For the first time in the history of the search for sentient life, telescopes can be pointed at stars known to host planetary systems, including those that may feature planets similar to Earth – just the type of worlds that might be home to a civilisation capable of building radio transmitters and receivers.

We assume that other intelligent civilisations would broadcast radio signals just as we do, but although radio-loud ourselves over such a short phase, we are already reducing our interference through switching to fibre-optic cables and telecommunication satellites. Furthermore, we are assuming that other intelligent civilisations would even want to talk to us, or know to look in our direction. What if other planets were host to hyper-intelligent space dolphins, who never rise above the sheltering waters of their world, in an attempt to evade dangerous stellar radiation? They would not even know there were other stars, let alone other intelligent beings wanting to communicate with them, nor would they probably care.

If listening for the one unexplainable radio message is a bit of a long shot, how else could we go about it? *Laser pulses* are a favourite choice. Russian and American scientists have scanned the skies periodically over the last couple of decades hunting for laser light, which is not only distinguishable from other natural types of light such as starlight, but as far as we know can only be produced by an intelligent source. Ghostly subatomic particles called *neutrinos* are perhaps better suited for transporting a message over long stellar distances than radio or optical signals, so it may be wise to look out for neutrino-based alien Morse code as well. We could also look for evidence of asteroid mining – humans are already seeing the potential benefits to our own civilisation of obtaining mineral resources from our local rocky belt, so why would an alien civilisation not do the same? Perhaps evidence could be found through unusual changes in the chemical composition of the asteroid belt, irregular-shaped chunks missing from images, an increase in the size and amount of debris surrounding a celestial body, or other changes detectable from Earth. An unfortunate side effect of our own culture is pollution, but this means we could search for similar dirty signatures in alien atmospheres. If there are non-natural chemicals, such as chlorofluorocarbons, in a planet's atmosphere, this would also be a sign that there might be someone with technology on the ground.

Taking this a step further, we could actually look for evidence that may have been sitting right here on earth for billions of years – who is to say that aliens have not been here at some point already and left behind some artefact or message for us to find? Our DNA encodes information – could it have an alien message written into it? It is of course highly unlikely, but still in the outer realms of possibility. We could even take a cue from science fiction and look for the signature from an alien spacecraft zooming by. We may just get lucky one day and receive an email. A group of scientists have set up a website inviting ET to get in touch, and although 99.9 per cent are hoaxes, only one needs to be real.

The Power of Civilisation

Believe it or not, should we ever make contact with ET there actually exists a ruler by which alien civilisations can be measured. It is called *The Kardashev Scale*. Currently classed on this scale by Carl Sagan as a Type 0.7 civilisation, the question being put is whether humanity will ever advance past the Moon and finally make our way into the ranks of the Type Is? As a civilisation grows and advances, its energy demands will increase rapidly as a result of population growth and the power requirements for the technology this greater number of people will supposedly build and operate. The Kardashev Scale was created to measure a theoretical civilisation's mechanical progression against how much energy it has available. The scale was originally designed in 1963 by the Russian astrophysicist

Nikolai Kardashev, who created three base classes, each with an energy disposal level: Type I (10^{16}W) Type II (10^{26}W), and Type III (10^{36}W). Recent astronomers have extended the scale to Type IV (10^{46}W) and Type V (all the energy available in all universes and in all realities). The human race still has a long way to go before being awarded Type I status as we continue to sustain our energy needs from fossilised plants and animals, and is therefore at the bottom of the civilisational heap.

A *Type I* designation is bestowed upon populations who have been able to harness all the energy available from their host planet and the energy that reaches it from their own star. The population will then have the knowledge and know-how to collect and stockpile this energy to match the demands of its growing numbers. However, humans being able to harness all Earth's energy is hard to fathom – it would mean we could control all the natural forces on the Earth, such as volcanoes, the weather and even earthquakes.

A *Type II* civilisation can harness the power of its local star directly, controlling the star itself. One hypothetical method of doing this is called a *Dyson Sphere*, named after the physicist Freeman Dyson, who had the idea that a growing technological culture would ultimately be limited by access to energy, and that advanced power-hungry civilisations would be driven to harvest all available light from stars. Such a device would consist of vast clusters of machines that would encircle a star, harvesting most or all of its energy output and transferring it to the civilisation's own planet for later use. Another idea for capturing stellar power involves the control of nuclear fusion, the mechanism that runs stars, and the harnessing of this power in an incredibly large reactor. Perhaps nearby gas giants could also be utilised for their hydrogen, slowly drained of their chemical life source by an orbiting nuclear reactor. What is attractive about this level of advancement is that should a moon-sized object enter the Solar System on a collision course with our little planet we would have the ability to vaporise it. Perhaps we could even play a planetary game of chess – moving our planet out of the way or sacrificing another planet of our choice to block the invader's path – cosmic checkmate.

Searching for evidence of this theory, at the end of 2015, scientists monitoring NASA's Kepler space telescope observed a bizarre star snappily entitled KIC 8462852 that was shown to be emitting a strange light pattern. As detailed in Chapter 8, Kepler searches for dips in starlight created when a planet transits in front of its host star. KIC 8462852 stood out as being strange because the star was witnessed dimming by 20 per cent, unlike the normal one per cent created by the passing of a planet. It was also seen to be surrounded by a mass of matter consistent with debris found around a young star soon after its formation; this star, however, is not young and the debris appeared to be recent. Could this debris in fact be a megastructure similar to a Dyson Sphere? SETI astronomers have so far found no evidence of radio signals coming from this potentially alien astro-engineering project, but KIC 8462852 will now undoubtedly be the subject of SETI observations for years to come.

Once a civilisation has graduated from gaining control over a planet and then a star, making its own extinction almost impossible, what could possibly be next? Well, a *Type III* civilisation, of course: a galactic space-roving population with total control of energy, resulting in a master race with dominion over the Universe. These beings may even have evolved into self-replicating cyborgs or cybernetic organisms, both biological and robotic. In this scenario, the descendants of regular humans would be seen as a weak, inferior and primitive clan. The dominant humanoid species would have a population boom as they colonised the Galaxy, building numerous Dyson Spheres to leach the energy from each new star they came across, creating a huge network of stellar energy funnelling power back to the home planet.

Not alone in his beliefs, Kardashev considered a *Type IV* civilisation too advanced to contemplate. Could anything be more theoretically advanced and fictional than Dr Who-esque cyborgs harnessing the power of hundreds of stars? Nonetheless, some believe that even greater advancement is possible for a society. Step forward the *Type IV* civilisations that would be capable of harnessing the energy content of almost the entire Universe and travelling across the accelerating expansion of space. A Type IV civilisation would need to tap into energy sources unknown to us using bizarre, or currently unknown, physics. There is a further level – a *Type V* civilisation, where beings would essentially be gods, with the knowledge and power to manipulate the Universe and beyond at their will.

Cosmic Sightseeing

The human brain is the most powerful entity in existence as we know it – a biological supercomputer. As such, it gives humans a massive advantage over automated machines such as telescopes, rovers and space probes. It has superior pattern-recognition capabilities and allows for ingenious problem-solving strategies. In space, the unexpected is the norm and no being is better than a human at responding quickly to the unanticipated. Despite the fact that we would be the best entities to send in search of other life forms, long-distance space exploration is highly problematic for the human body. If we ignore the obvious issue of a lack of advanced-enough technology and propulsion systems even to blast humans off to far-flung star systems, our weak little bodies would not survive the trip. A short lifespan of up to 120 years is not compatible with the up to 80,000 years it would require with current technology to reach even the closest star, not to mention what spending that length of time in microgravity and exposed to the radiation of space would wreak physically on the human body and mind. It represents a perfect storm of immeasurable distance, slow travel speed, human frailty, short lifespans, and extreme cost, which in combination necessitate our reliance on technology to reach out to the stars and make the introductions for us.

This is not a bad second choice. Automated robotic probes have been sent to planets, moons, comets and asteroids, and to the very edges of our Solar System, and for the most part have been successful beyond our wildest imaginations. However, what about going beyond the Solar System to search for other life forms? That is a goal far beyond our current capabilities. In 1978, the probes *Pioneer 10* and *11*, were launched and after successfully completing their missions around Jupiter and Saturn, continued along paths that would eventually take them out of the Solar System. Today, they are heading towards the star Aldebaran, 68 light years from the Sun (which will take in the region of 2 million years). After 30 years of flight, they are still somewhere well within the Solar System and sadly all radio communications have been lost. Even though we will never know what, or even who, these probes meet on their journey, *Pioneer 10* does bear a plaque inscribed with information about Earth and humanity – just in case it bumps into some beings from the Aldebaran civilisation. But if they reply to our message, will there still be anyone left on Earth to receive it?

Since voyaging into the cosmos looking for friends is challenging, and two-way communication is not possible with our current technology and short lives, we are focusing instead on sending *messages in bottles*, as well as listening out for them calling to us. Alien civilisations may even be able to eavesdrop on humanity by tapping into our TV and radio broadcasts. Considering that one of the first TV transmissions was of Hitler opening the 1936 Olympic Games – and we all know what happened soon afterwards – a worry is that this will not exactly inspire them to get in contact. In 1951, the first episode of *I Love Lucy* was broadcast and some 0.0002 seconds later, the signal headed into space. Maybe this, or re-runs of *Friends* or David Attenborough documentaries could make humanity and the Earth seem more appealing.

Given that stars in our galactic neighbourhood are separated by about 4 light years, in the past 50 years roughly 10,000 star systems may have been exposed to our TV shows. They must be rather confused about what on <insert alien planet name here> is going on over on that little pale blue dot – or perhaps they are as hooked on *Game of Thrones* as we are.

Hello?

We are not only broadcasting BBC Radio 2 and reality TV shows, but have actually sent complete messages into space. SETI most famously transmitted a communication to the stars in 1974 using radio waves from the *Arecibo* telescope in Puerto Rico, aimed at a globular cluster, M13, more than 25,000 light years away. Frank Drake and Carl Sagan composed the message, which included the numbers 1 to 10, atomic numbers of the elements of life such as hydrogen, carbon and oxygen, information on our DNA, a figure of a human (non-gender specific) and a graphic of our Solar System highlighting the Earth as the origin of the message. Sadly, the stars this message was aimed at will no longer be in the same spot by the time it arrives, but who knows who might pick it up along the way.

Three years later in 1977, the *Golden Records* were launched on both *Voyagers 1* and *2*, intended to communicate a story of our world to extraterrestrials, interestingly only portraying the positive sides of Earth – no warfare, hunger or disease – which makes sense. Who wants to make first contact with an alien civilisation only to highlight the less attractive aspects of life *back home*? In August 2012, *Voyager 1* took its first steps towards becoming that beacon as it entered interstellar space, leaving our Solar System behind. The spacecraft is trekking towards a star called Gliese 445, and has a date with it in 40,000 years' time. Of course, by this date it won't be able to transmit home any data – in fact, by 2025 all of its scientific equipment will have stopped working (the equipment on board is already more than

40 years old, created before the CD and colour television). The *Voyager* message, however, is embedded in a 12-inch gold-plated copper disk containing sounds and images selected to portray the diversity of life and culture on Earth. Again, Carl Sagan led the charge and with his associates assembled 115 images and a variety of natural sounds such as surf, wind and thunder, birds, whales and other animals. To this they added musical selections from different cultures and eras, and spoken greetings from Earthlings in 55 languages, together with printed messages from the then President Jimmy Carter and UN Secretary General Kurt Waldheim. The launching of this record says a great deal about the hope and positivity of humanity. Since then, all missions to other planets and those designed to orbit and explore the Solar System have contained messages including information about the Earth and humanity, just in case any being comes across them.

After *New Horizons* flew by Pluto in 2015, it began a new journey towards the Kuiper Belt, following which it may one day become the fifth spacecraft to leave the Solar System and may even be the first to be discovered by an alien species some millions of years from now. Unlike *Pioneers 10* and *11* and *Voyagers 1* and *2*, *New Horizons* was launched in 2006 without a welcome message, but following completion of its active mission there will be some space available on its computer for a message to be uploaded digitally. Unlike previous messages, however, this one, fondly known as the *One Earth Message*, will be a unique crowd-sourced message in a bottle, sent by people from all over the world. Who speaks for Earth? The answer is *everyone*. This literal selfie of our planet will aim to communicate to ET the real essence of the Earth, humanity and the other life forms that share our planet. If we did not believe there was even the slightest chance of intelligent life out there, why would we bother?

Much of today's scientific exploration and data analysis looks to the public for help and support, such as the

Zooniverse, which calls upon citizen scientists to help comb through mountains of data to aid scientists with classifying distant galaxies, spotting black holes, characterising surface features on Mars and even hunting for exoplanets. The quest for ET is no different. The SETI@home project has involved the worldwide public in a search for radio-wave evidence of life outside Earth for 16 years. Based at the Space Science Laboratory at the University of California (Berkeley), this project records and analyses data from the Arecibo Observatory by searching for narrow-band signals of possible extraterrestrial origin. As yet, no such signals have been found. Today, SETI@home continues its search for evidence of extraterrestrial life with hundreds of thousands of volunteers each hoping to be the one to find it.

What Happens if We Find it? What Happens if We Don't?

As Carl Sagan wrote in *Contact*, 'The universe is a pretty big place. It's bigger than anything anyone has ever dreamed of before. So if it's just us ... seems like an awful waste of space.'

NASA's Kepler space telescope has helped scientists discover thousands of exoplanets, and has a very large field of view of 105 square degrees – comparable to the area of your hand stretched at arm's length. Most astronomical telescopes have fields of view of less than 1 square degree so, although Kepler can monitor more than 100,000 stars, it is still a miniscule area of the Galaxy, let alone the Universe. Set to launch in 2018, NASA's next-generation James Webb Space Telescope, together with its larger successors, will give scientists the opportunity to look for signatures of life in the atmospheres of exoplanets – although they will not be capable of distinguishing whether life forms are brainy beings or single-celled microbes. Astronomers now know that every star in the Milky Way galaxy has at least one planet orbiting it, so humanity's first contact with alien life may one day be possible. It is also not just a case of where to look, but also when. The Arizona State University astronomer and author Paul Davies points out that even if a fairly close civilisation, say one 1,000 light years away, were to look through a telescope and find Earth, it would see the planet 1,000 years in our past. Why would they bother to send a message to a planet that had not yet discovered electricity, let alone built a receiver to intercept their message?

Even the discovery of some simple sort of ET life would be extraordinary. Finding non-sentient extraterrestrial life would help piece together our own origins and the history of life on Earth, and would be a momentous watershed event for the entire globe. If the discovery were of life that could say *hello* back to us – life as we know it would change forever.

UFO-spotters, Raëlian cultists and self-certified alien abductees notwithstanding, humans have to date seen no sign of any extraterrestrial intelligent civilisation. We have not received any visitors from space, nor have our radio telescopes detected any unusual transmissions from other worlds. But although no one has yet found life elsewhere, there's no reason necessarily to despair. Mass extinctions have wiped out vast majorities of species in our planet's nearly 5-billion-year history, and yet here we are. We can only assume and hope that any life existing elsewhere would be just as resilient.

But what if this really is all there is – and our isolation extends far into the Universe, or, and perhaps more likely, we do not recognise different life forms because some of our presumptions concerning how an alien civilisation might look and behave, based on our own experience, are incorrect? In 2015, the close examination of 100,000 galaxies near to our own concluded that none presented any irrefutable evidence of civilisations with highly advanced technology. Instead of listening for voices from the skies, scientists looked instead for heat signatures that would be produced by advanced civilisations, just as Freeman Dyson and later the Kardashev Scale predicted might exist. The idea is that, once a star is encased in a Dyson Sphere, its glow would be suppressed, but the engineered construction itself could be detected by the wasted heat oozing out from it. The same process is happening when your iPhone warms up during prolonged 'Googling'. In some sense it doesn't matter by what means a galactic civilisation generates or uses its power, because the second law of thermodynamics makes energy use hard to hide. They could construct Dyson spheres, draw power from rotating black holes or build giant computer networks in the cold outskirts of galaxies. Any of these would produce waste heat, but the search for objects emitting more heat than light in over 100,000 nearby galaxies has yielded, perhaps unsurprisingly, no positive results. No Type III civilisations have been found – yet.

Over the years, to explain away the endless silence of deep space, researchers have created a vast assemblage of possible explanations for the disappointing lack of any intelligent alien life. Perhaps we are indeed alone. Perhaps the laws of astrophysics and biology make intelligent life vanishingly rare, and the rise of humans was serendipitous. Perhaps technological civilisations always destroy themselves once they reach a certain point. Perhaps interstellar travel is simply too hard, too slow or too boring for any advanced civilisation to bother undertaking. It could be that galaxy-sterilising explosions such as gamma-ray bursts in the cosmic past suppressed the rise of advanced civilisations and now that these have quietened down, we humans had the chance to arise and are the foundations for civilisations yet to come. Perhaps, and I like this idea best, any advanced civilisation will have become in tune with its natural environment, value nature's role in its survival and will be working in harmony with it. In this scenario, our supremely intelligent and compassionate aliens would not produce waste heat, light, nor electromagnetic signals from a repository of profligate technologies – so it will be less easy to find them until we use our intelligence to do the same.

CHAPTER TEN

The Next Generation

Imagine the Earth in the future. Are there flying cars, floating cities, outposts scattered across the ocean floors, and artificially intelligent cyborgs? This is the magical, technologically advanced world that science fiction paints. But is it realistic? By 2050, the UN predicts that Earth's human population will have grown from 7.1 billion to between 8.3 and 10.9 billion and may only continue to grow. In this version of the future, many questions will surround the sustainability of world populations, the growing pressures on the environment, global food supplies and energy resources. Will the Earth be able to sustain us? With the answer to this question unknown but worrying nonetheless, humanity needs to start planning for the future and to contemplate leaving the safety net of the Earth and looking towards the stars. Stephen Hawking has stated that the colonisation of space would be the best way to ensure the survival of humanity. Although I believe there are other ways to ensure our species lives on, space settlement is the next logical step after space exploration. But if we leave our home, suddenly we will be living outside of what we consider to be 'normal' and will be attempting to inhabit environments in which we are not originally biologically designed to survive. We will have to become the extremophiles, a generation of aliens and Life 2.0.

Why and Where Should We Go?

When in 1961 Russian cosmonaut Yuri Gagarin was blasted into orbit and safely returned, he became the first human in space and people were finally given substantial evidence to support the idea that humans could travel off-world. Subsequently, and since the year 2000, humanity has been continuously living in space. Our permanent cosmic residence is helping us prepare for a future when humans may need to be able to live and work on other planets and moons. Although a colony of fewer than 10 people, the International Space Station is our first extraterrestrial outpost, an inhabited satellite orbiting the Earth. No terrestrial or land-based space colonies have been built thus far, yet this is the dream, as investigating the habitability of other worlds not only benefits our understanding of the versatility of life, it also takes us closer to answering one of humanity's oldest questions – are we alone?

In the long term, *i.e.* another 3 billion years, our Sun will start to expand and enter its red-giant phase as it draws closer to its death. It will engulf Venus, and even if it doesn't swell enough to reach the Earth, it will still boil off the oceans and heat the surface to temperatures that even the hardiest extremophile could not survive. This, however, is a rather long way off and not really a good enough reason to start the process of moving right now. There is no denying that it would be magnificent to have people living on multiple worlds, nor that sadly the Earth is beginning to sag under the pressure of humanity. We are on the verge of self-inflicted destruction, whereby the damage we are doing to our planet is progressing faster than our capabilities to fix it. Off-world colonies could, and I strongly emphasise *could*, improve the chances of human civilisation surviving in the event that the Earth becomes uninhabitable. Life is fragile and any number of natural or man-made catastrophes could occur, such as another Snowball Earth, asteroid impact, nuclear war or complete depletion of our natural resources. Hopefully long before any of these scenarios come to pass, I believe we will choose to leave Earth for the purpose of exploration and scientific enlightenment, because it is built into us to want to travel and unravel the mysteries beyond our physical and intellectual borders. If we happen to set up humanity's lifeboat in the process, then even better.

As such we need to explore how humans might be able to live and work on other planets, moons and in space itself. What might it be like to live on other worlds in our Solar System? A great thought experiment with a similar theme and plot-line to the film *Interstellar*, assumes that we have developed the capability to skip across the Solar System to its farthest reaches and have the knowledge and technology needed to build a human outpost on any world of our choosing in the Solar System. So where would we go?

Mercury

Starting from the inside out, humanity could move to Mercury – and it goes without saying, its surface would be an extremely inhospitable place in which to land unprotected. It would not be our first choice, that's for sure!

Like the Moon, it lacks a protective atmosphere, so colonists and their equipment would need thermal protection from the intense heat of the Sun, and require shielding from the powerful solar radiation that reaches the surface and infrared radiation of any very hot region of Mercury's crust. Because of its rocky, barren and airless similarities with the Moon, any settlement of Mercury might be performed using the same general technology, approach and equipment as a colonisation attempt on the Moon. Unlike the Moon, however, Mercury has the advantage of a magnetic field that protects it from cosmic rays and solar flares, and a larger surface gravity of about 0.37g, almost exactly equal to that of Mars – just over one-third of the gravity on Earth. This means that heavy equipment and building materials would be easier to lift and move around and, as an amusing bonus, a human could jump three times as high. Due to its proximity to the Sun, the surface of Mercury can reach 427°C (800°F) near the equator during the day (hot enough to melt lead) and fall to −180°C (−292°F) at night, with temperatures at the poles being even colder – that's a lot for the human body to deal with. For a planet located so close to the Sun, it is perhaps surprising that significant deposits of ice lie hidden in the shadows of polar impact craters, but could this ice be mined to access water on Mercury? These polar regions could actually be a good spot for settlements, providing a source of water and a break from the intense heat of the Sun. Weirdly, a colony would not really have to worry about any natural disasters wiping them out as, without an atmosphere, Mercury has no weather and with no bodies of liquid water or active volcanoes, there is very little risk of devastating tsunamis or volcanic eruptions. This lack of atmosphere would also mean that during the day, the sky would appear black, not blue – nothing to worry about, it would just be odd! Sadly, like all planetary bodies in the Solar System, however, there is always the threat of asteroid impacts and, potentially, earthquakes (Merquakes?!), owing to compressive forces that shrink and squash the planet.

Happily, it only takes around five minutes for signals to bounce between Mercury and Earth, so there could be a stunted but very achievable conversation with home. Who would have thought that Mercury would actually be quite an attractive option?

Venus

Next for consideration is Venus, our Solar System's equivalent to the mythical realm of Hell. Already it sounds appealing, doesn't it? Without technological help (which is far beyond our capabilities at the present time), our fragile human bodies would die in less than 10 seconds on the surface, instantly crushed while being simultaneously cremated. The last, very quick, breath taken would be of toxic gas. Life on Venus would be nauseating, brutal and very short. If we could somehow surmount these issues, then the best place to set up camp would be somewhere on the flat, smooth plains that can be found on more than two-thirds of the planet. Walking around would not be a pleasant experience with sweltering surface temperatures and air so thick that every step would be like trying to run in water. The planet's gravity would not be a problem, however, as it is almost 91 per cent of that of Earth so would probably not feel much different, but the atmospheric pressure is 92 bars – tantamount to living more than 900m (3,000ft) beneath the ocean on Earth.

High in the atmosphere on Venus, winds travel up to 400kph (249mph) – faster than any tornado or hurricane witnessed on Earth – interspersed with fierce bursts of lightning that, as a small mercy, never reach the already challenging surface. The roasting heat prevents any rain from touching ground so water is never able to collect or even wet the surface. The active volcanoes on Venus may add yet another danger to colonists as the eruptions are thought to be so large that they could re-surface the whole planet. It would take only a few minutes to get a distress call home when the planets are closest, but when Venus disappears to the other side of the Sun it would take up to 15 minutes. Not that any help would be close at hand should it be needed, nor could anyone really offer assistance against such an inhospitable surface environment.

The type of habitat that could use the strengths of Venus' extreme conditions is one filled with gases of the same composition as Earth's atmosphere at sea level (so colonists could breathe) but floating high in the dense Venusian atmosphere. The atmospheric pressure at 50km (31 miles) above the surface of Venus is similar to that on Earth at sea level (1 bar), and temperatures are just over 0°C (32°F) at that altitude. Just like a weather balloon, a floating habitat with internal pressure like that on Earth would rise to an altitude on Venus where the external pressure was the same. If this were actually possible, a human could walk outside on to a ramp using just an oxygen tank and look out over the Venusian clouds below. The habitats would produce their own oxygen through photosynthesising vegetation, which would not be difficult to grow in Venus's carbon dioxide-rich atmosphere, and water could be extracted from the sulphuric acid in the clouds. Theoretically, anything is possible.

The Asteroid Belt

Although perhaps a surprising choice, there is a real option of moving to a dwarf planet such as Ceres, the largest object in the Asteroid Belt. The objects of this orbiting rock garden have been suggested as potential sites for future space-mining operations hoping to reap water for its hydrogen to make rocket fuel; oxygen to provide breathable air to make long-distance space missions possible; and ore minerals. They are also seen as potential staging posts and transport hubs for deeper space exploration due to their lower escape velocity. The main fear any colonists would have living on these objects would be of one asteroid bumping into another, and the resulting threat of knocking objects out of the belt and on to a collision course with Earth.

Comprising one-third of the mass of the entire asteroid belt, Ceres could quite possibly become the main asteroid base for future trips to Mars, as it is not a dead lump of pockmarked rock, as commonly comes to mind when we think of an asteroid. Ceres may actually contain more water ice buried beneath its surface than all the fresh water flowing across the Earth, and although its gravity is less than 3 per cent that of the Earth, it is one of the most suitable locations for a permanent human base. Ceres is currently not believed to have a magnetic field so its surface is not shielded from cosmic rays or other forms of radiation, and it does not have a significant atmosphere, so there is no weather (adverse or otherwise) and it always remains well below freezing. NASA's *Dawn* spacecraft was the first to visit Ceres, having arrived in orbit in March 2015, and was greeted by a complex and beautiful landscape full of weird-shaped craters, a 6.5km- (4-mile-) high mountain, mysterious bright spots that might be huge deposits of salt, and an entirely dry surface. Excitingly for life, however, there may be liquid water inside the dwarf world as water vapour has been seen erupting from it into space, possibly from volcano-like icy geysers. Thankfully, these vapour jets would be far too weak to pose any danger to humans but they hint at the world's potential to support life. The relatively small size of Ceres is a selling point as well. Hiking across its surface to find a suitable site to live would not take very long (relative to travelling around a planet or moon that is) – Ceres has a diameter of 950km (590 miles), just short of the distance from the south coast of England to the north of Scotland.

Europa

Despite depictions within some entertaining science-fiction films, humanity could not even contemplate living on Jupiter. A purely gaseous world, there is nowhere for a crew to touch down, which makes colonisation impossible – unless you fancy setting foot directly on to its core and can tackle the crushing weight of liquid hydrogen bearing down on you. We might be able to live in orbit around Jupiter, however, and harvest its energy to provide the resources for colonisation of any number of its 60 nearby moons. If you ask an astrobiologist they will instantly suggest setting up a base on Europa due to the tantalising science we could accomplish, whereas a geologist might suggest heading to Io to study the many active volcanoes. In reality, Callisto with its own large amounts of water ice, low radiation levels and relative geological stability would be ideal. However, where is the adventure in that? For human explorers setting up a research base on Europa, the cold, icy surface would actually be quite suitable. It is relatively flat and, although crisscrossed with small ridges, these are little more than a few metres high so should not prevent construction nor journeys across the landscape. A serious threat to life, however, is Jupiter's magnetosphere, which bombards Europa with deadly radiation. The best location for a base would therefore be either protected below Europa's icy crust or on the hemisphere of the moon that faces away from Jupiter, as this receives the least amount of radiation. Like our Moon, gravity is low (about 13 per cent that of the Earth), so going for a walk on either moon would be a similar experience, and also means that both have an almost imaginary weatherless atmosphere. Colonists, if daring to live on Europa's surface in inflatable igloos, would need them heated to help deal with the blistering cold outside (down to −220°C/−364°F at the poles), reinforced to withstand icequakes, and located away from possible powerful plumes of water shooting out through the icy surface from layers far beneath. If you wanted to email research findings home, a message would take at least 30 minutes to arrive, and only when the gargantuan bulk of Jupiter wasn't blocking the way. Oh, and don't even think about making a voice call.

Enceladus and Titan

For the same reasons as Jupiter, Saturn is not a place where you want to end up. However, if you had to make an emergency landing somewhere in this system and could make it to Titan, then you might have a chance of survival as you could jump out of your ship without the need for a pressurised spacesuit – you'd simply need a tank of oxygen and some insulating clothing. With its thick atmosphere, standing on the surface of Titan would feel rather like being submerged in a swimming pool on Earth. The landscape also resembles that of the Earth and there is a number of flat areas of land for a colony to build upon. If this weren't enough good news, Titan is primarily composed of water ice and rocky material and NASA's *Cassini* mission showed exciting hints of an ocean inside it, which might be as salty as the Earth's Dead Sea. Colonists would just need to melt this surface ice, and/or access the ocean and filter out the salt. One downside is that although the surface is covered in aqueous bodies, humans could not drink the hydrocarbon-rich liquid. Excitingly, Titan offers a lot to work with, as it already possesses an abundance of all the elements necessary to support life. Water can easily be used to generate oxygen and nitrogen to add to breathable air. Nitrogen, methane and ammonia can all be used to produce fertiliser for growing food. The best thing about Titan concerns transportation, which takes on a whole new dimension because of the moon's low gravity (more or less 14 per cent that of Earth) and dense atmosphere – colonists could strap on wings and fly! It has weather in the form of methane rain and thunderstorms but no cyclones or tornadoes, and its atmosphere would protect colonists from cosmic rays and many incoming projectiles. In addition, there are no moonquakes as far as is known and the existence of cryovolcanoes is still debated – it is a pretty comfortable, safe and quite possibly enjoyable place in which to live. It gets my vote!

If a spacecraft carrying colonists ended up on Enceladus, one of Saturn's other moons, then life would be a lot harder. The main dangers here are the freezing cold temperatures, minimal air pressure and explosive geysers. Due to the moon's icy surface, most of the sunlight it receives is reflected back into space, lowering the temperature to an average –201°C (–330°F) throughout the day. As a result of its sparse atmosphere, it has no extreme weather for colonists to worry about, but is left exposed to incoming space debris and radiation. The best locations for a human base would be near to the *tiger stripes* in the southern polar region as these would provide a source of heat and power. These giant fissures are where the plumes of frozen ice particles and cold vapour spew out into space, and produce nearly 16 gigawatts of power – a reasonable trade-off for the danger of living close by. However, the moon's tiny gravity, just 1 per cent that of the Earth, would hamper travelling around. It is not really a world you want to end up on forever.

Uranus and Neptune

Once we get to this frigid neck of the Solar System, colonisation is much more possible on moons than planets. This is because, just as with Jupiter and Saturn, Uranus and Neptune lack much in the way of a solid surface under their layers of ice and gas on which to settle. The pressures below the thick atmosphere on Uranus are enormous and would instantaneously crush any life form. Also, there is no process inside Uranus, such as volcanism on Earth, that would give colonists a form of energy to use as a replacement for the very distant Sun. Out of the 27 moons orbiting Uranus, two prime targets for colonisation would be *Titania* and *Miranda*. Not a huge amount is known yet about these moons, but it is thought that they have a solid surface on which a mission could at least attempt to touch down. All of Uranus's moons lack weather systems and surface pressure due to non-existent atmospheres, and it is probably a safe assumption to say there is a multitude of as yet unknown hazards waiting on them. They are also very cold, with the average temperature of Titania, for example, hovering around −203°C (−330°F). Furthermore, all the moons spend 42 years in darkness and 42 years in faint sunlight, which would not exactly prove ideal for the human body or mind.

Humanity could also happily bypass Neptune and move on to its largest moon, *Triton*. Little is known about this moon either, but we do know it is made of rock and nitrogen ice and has both cratered and smooth regions. The smooth areas are formed when geysers of dust and nitrogen gas erupt out of the moon's crust. The dust then drifts gently back down, coating the surface of the moon. It has a slight atmosphere and might feel strangely similar to standing on the Moon or Mercury. It's unclear, however, how dangerous the geysers would be but, as with any unpredictable eruption, establishing a settlement next to or near one would never be a smart move. Triton is currently the coldest known object in the Solar System with an average temperature of −235°C (−390°F), so a continuous artificial energy source would be needed to keep a colony warm.

Pluto

This dwarf planet has been shrouded in mystery for so long that science can only speculate about what setting foot on it might be like. Depending upon where it is on its 248 Earth-year orbit, freezing temperatures can be expected down to −233°C (−387°F), and so we can kiss goodbye to the chance of liquid water. It has a tenuous atmosphere, created through the seasonal sublimation of ices on the surface, but is not thick enough to give the surface much pressure to work with – just 0.003atm (0.3 MPa/3mbar) – and due to its small size has only 1/15th the gravity of the Earth. However, NASA's *New Horizons* probe has given Pluto a much-needed confidence boost (it was owed nothing less after the whole 'not a *real* planet' demotion) by revealing an incredibly geologically diverse and active world, even out in the farthest reaches of the Solar System. Should humanity ever figure out how to travel there in person, Pluto would display a number of useful attributes. This newly viewed second Red Planet is reddish owing to layers of haze stretching 160km (100 miles) into the atmosphere, while the surface is covered in flowing nitrogen-rich ice, similar to the movement of glaciers on the Earth, as well as ice volcanoes and snowfall. There may even be an underground ocean. Although setting up and maintaining a settlement on Pluto would be enormously complex and communications would be frustratingly slow, it would, scientifically at least, be well worth the endeavour.

Living in Space

At our current level of technology, the building of a space colony would present a set of extraordinarily great challenges. Space settlements would have to provide for all the material needs of hundreds or thousands of people, in an environment that would be very hostile to human life. Colonists would require systems such as controlled life support, which have yet to be developed in any meaningful way, and would be obliged to deal with isolation and confinement over many years, potentially for the entirety of their lives. The huge cost of sending anything from the surface of the Earth into orbit (roughly £15,000 per kilogram) gives an insight into the astonishing costs associated with building and launching a space colony.

As mentioned previously, humanity is already living in space and life on the International Space Station (ISS) provides a glimpse into some of the major challenges humans would face should we venture further into the Solar System. The *ISS* is a habitable satellite that orbits the Earth at an altitude of 355km (220 miles) once every 90 minutes, meaning that the Sun sets and rises for the crew nearly 16 times a day. It's a vast project with shared ownership by NASA (USA), Roscosmos (Russia), JAXA (Japan), ESA (several European countries) and CSA (Canada), who all pitched in to build it. For the last 15 years there have been up to 10 astronauts at any one moment living in the vacuum of space above our heads, for up to a year at a stretch. Astronauts from all contributing space agencies have spent time there and the first British–ESA astronaut, Tim Peake, arrived for a six-month getaway in December 2015.

There are so many scientific and spiritual benefits of spending time in space, but it is the challenges and dangers that these real-life superheroes overcome that tend to capture the public's imagination and most certainly deserve our respect. The best-known attributes of the cosmic environment are that there is no oxygen or pressure in the vacuum of space. Daring to take an unprotected breath in space removes oxygen from the blood without replenishing it, so after 9–12 seconds, the deoxygenated blood would reach the brain, resulting in loss of consciousness. Two minutes later, death would follow. Blood and other bodily fluids would boil as the pressure instantly dropped, causing the body to swell to twice its normal size, but it would not explode, as commonly depicted in films – this is a myth. Another myth to debunk is that portrayed by the image of a frozen drifting corpse. In the vacuum of space, there is no medium for removing heat from the body, so in fact an astronaut is very unlikely to freeze to death. A vacuum flask is exceedingly good at insulation and keeping coffee hot, and it would also be true of your body warmth in space. Rapid evaporative cooling of skin moisture in a vacuum may create frost but this in itself is not fatal.

Without the protection of Earth's atmosphere and magnetosphere, astronauts are exposed to high levels of radiation. A year in low-Earth orbit results in a dose of radiation 10 times that of the annual dose on Earth, which damages the lymphocytes in the blood – cells that are heavily involved in maintaining the immune system – and DNA itself. This damage contributes to the lowered immunity experienced by astronauts and potentially gives them a slightly higher risk of developing cancer later on in life. Thankfully, the crews living on the *ISS* are partially protected from the space environment as they are in a low enough orbit still to be embraced by Earth's magnetic field, which deflects the solar wind around the Earth and the *ISS*. Nevertheless, a solar flare ejected from the Sun is still powerful enough to warp and penetrate these magnetic defences, and be hazardous to the health of the crew. Beyond the limited protection of Earth's magnetosphere, however, interplanetary manned missions are much more vulnerable.

The greatest challenge, other than funding, facing human space exploration is not the technology required to achieve it, but the fragility of the human body to withstand it. To survive for a prolonged or even indefinite period of time in space, the effects and impact of long-term space travel on the human body must be properly understood, and for that there needs to be some voluntary test subjects – the astronauts. Technology has proven its ability to shield their bodies from many of the dangers of space, either by creating a life-support system to provide air, water and food and maintain comfortable temperatures and pressures, or by building a spaceship hull and habitat for shelter and protection against hazardous radiation and incoming micrometeorites. One aspect of a space-based life that cannot be avoided or protected against, however, is that of microgravity. If you have ever ridden a roller coaster and felt your body rise up as you crested the first huge hill and then plummeted towards the ground, you have experienced weightlessness. Imagine that feeling for maybe an entire year – no wonder over 40 per cent of astronauts feel nauseous. One astronaut, Jake Garn, was so unwell with space sickness that a new unit of measurement was named after him. The Garn Scale is now used as a gauge of how space-sick an astronaut is – the top level indicates when he or she just wants to give up and go home. What is surprising, however, is that the human body manages to adapt remarkably well to living in zero-g or, more precisely, microgravity. But the effects go far beyond the initial trip. Temporarily, weightlessness causes many key systems of the body to relax, as there is no longer the need to work against the pull of gravity. Astronauts experience disorientation as their sense of up and down becomes confused, which is why the ISS has all of its writing on the walls pointing in the same direction. Also, ISS occupants have reported losing track of where their limbs are and of feeling as if they are not there anymore. Spacecraft design takes into account all the effects of microgravity by putting extra foot- and handholds everywhere. Some materials – including human facial hair – tend to be more flammable in lower gravity. Thankfully, handling hazardous combustible materials on the ISS is taken very seriously and carried out with great care, so the risk of astronauts burning off their eyebrows is pretty low.

The longer astronauts spend in space, however, the greater the enduring impact the lack of gravity has on their bodies. Most famously, they experience deterioration of bone mass. The calcium in their bones oozes out through their urine, weakening the bones over time and simulating accelerated osteoporosis. This condition is thankfully mostly reversible once back on solid ground and in Earth's gravity. Consequently, astronauts are much more susceptible to breaking their compromised bones should they slip and fall (those extra handholds come in useful). Sadly, an astronaut's muscles also lose mass because while floating around is all very pleasant, a space traveller would literally waste away if that were all he or she did. Although astronauts have to exercise for two hours a day in orbit in an effort to counteract this muscle-wasting, they still require months of rehabilitation to build muscle back up again once they have returned to Earth. Astronauts also grow an inch taller while in space owing to their spine elongating, and they can develop a swollen *moon-face* as the body's fluids move upwards. Unfortunately, this shift in fluids can also cause eyesight problems in astronauts, defined mainly by their seeing flashes and streaks of light. Much of this can gradually be reversed once back on Earth – but what if an astronaut were not returning to Earth? What would happen to the body then? The first one-year mission to the *ISS* launched in March 2015 conducted a unique experiment as US astronaut Scott Kelly has a twin brother, Mark, a retired astronaut himself, who remained on Earth. The brothers are now being studied to observe the effects on Scott's body in long-term weightlessness versus his brother's on Earth.

A Space Oddity

Life in the cosmos also means dealing with a very distinct lack of personal space. Best-known of the challenges facing astronauts are long-term isolation, monotony, limited mobility and living in extremely close quarters with the same small group of people. The ISS is vastly larger than any previous space structure, about the size of a five- or six-bedroom house, but even so, staying inside your house for six months is hard to cope with both mentally and physically. Astronauts have cramped living quarters, privacy is a luxury, and they have to share everything with their fellow crew members for months at a time. The constant noise of people and machinery and the irregular light patterns make it difficult to sleep on board the space habitat, with astronauts commonly experiencing fewer hours of regular sleep and/or poor-quality snoozing. Combine that with the disruptions of the natural Earth day/ night cycles en route and the result is stressed and fatigued personnel. Maintaining Earth standards of personal hygiene is also almost impossible as water is precious and showering in microgravity is not an option – but apparently this is a minor irritation and the *ISS* actually doesn't smell too badly.

Surprisingly, although astronauts are physically removed from the Earth, they actually experience less isolation than scientists living in Antarctica in the height of winter. Regular contact with Earth, chatting with mission control, family and friends, as well as surprise calls from celebrities, via both video and voice chat and email keeps these space-dwellers thinking positively and feeling connected, giving them a respite from day-to-day chores and providing a sense of comfort and normality in an alien environment. The Internet and the advent of blogging and Tweeting, for example, may also ease the feelings of isolation faced in space, knowing that in cyberspace there is always someone listening. For five months, from December 2012 to May 2013, Canadian astronaut Chris Hadfield served as the commander of the *ISS* and gained a reputation as the 'most social media savvy astronaut' by sharing his daily life with the world, posting over 45,000 photos on Tumblr and Twitter and recording videos for YouTube. His guitar-playing and vocal performance of the late David Bowie's *Space Oddity* and exchange of tweets with William Shatner of *Star Trek* legend were remarkable, considering it all was sent from space. For the first time, the *ISS* felt like simply an extension of the Earth. Hadfield himself said that posting the photos, and the immediacy of the reactions and collective sense of wonder he could share with people from all over the world, made him feel connected with the planet and to other people, even as he floated hundreds of kilometres above them. All those who have lived in space say they took great comfort in the view of and in communication with the Earth, but what if the journey meant that Earth became barely a dot on the horizon and communications nearly impossible?

Life in space has the potential to lead to depression, interpersonal conflict, anxiety, insomnia and even psychosis. Astronauts are living a life of risk. One rogue meteor or solar flare and it's all over. But the considerable preparations made by astronauts aim to help them fight any negativity caused by months of living in fear and isolation. They train together for years as a team and as a family, so they already have camaraderie, a mutual understanding and trust, and to some extent intimacy with each other. Since outer space is considered an extreme environment, most training simulations and camps are also located in remote and harsh environments. The NEEMO mission sends 'aquanauts' to the underwater *Aquarius* research station off the Florida Keys and several other analogue missions have been conducted on the Earth to simulate living in space or indeed on Mars. Most of these are research-based and were described in Chapter 5, but one – Hi-SEAS – is focused on the daily lives of people living off-world. Hi-SEAS (Hawaii Space Exploration Analog and Simulation) is a self-contained habitat found at an elevation of around 2,590m (8,500ft) on the slopes of Mauna Loa volcano on the Big Island of Hawaii. 'Terranauts' live in a geodesic dome simulating life in a close-knit colony on Mars, including communication delays, isolation, cramped living quarters and, most importantly, food preparation. It's no secret that pre-packaged, dehydrated space food is bland and the extent of seasoning involves pepper suspended in olive oil to stop it flying up and scattering around the station. Eating the same foods day in, day out can cause a syndrome known as menu fatigue, a common affliction at the *ISS*. The gastronomically bored astronauts end up consuming fewer calories, and ultimately lose weight and can become malnourished. NASA continually examines the daily lives of the crew on the *ISS* to see how they're coping in a harsh and isolated environment, and to improve the agency's plans for future long-term missions in space. Despite all the risks, there is no shortage of applicants for astronaut positions and virtually everyone who has had the chance to live in space is keen to return.

Fly Me to the Moon

We mentioned before the bodies in the Solar System where humans might one day make footfall, or at least how they might accomplish it should technology allow, but skipped over our nearest and dearest celestial relative, the Moon. Designs for and ideas about how humans might live on the Moon have existed since long before the dawn of the Space Age, but what is actually feasible today? And why the Moon?

The Moon is an ideal staging post where we can accumulate materials, equipment and personnel outside the confines of Earth's gravitational pull, and it could be used as a test bed for the technologies needed to place humans on other worlds. From the Moon, we can send missions onwards to Mars or into deep space, set up astronomical observatories to view the cosmos without the interference of an atmosphere or Earth's radio chatter, utilise lunar resources (mining deposits such as titanium and helium-3) and even support a bustling space tourism industry – who wouldn't want to take a weekend break on the Moon? Humanity already has the means to get there and there are technologies that have proven advanced enough to sustain human and plant life in space. We just need the commitment and finance to proceed with it.

To build a habitat on the Moon is no easy feat. It requires consideration of how building materials will respond to the Moon's vacuum; the extreme temperature variations between day (120°C/248°F) and night (down to −153°C/−243°F); impacts by micrometeorites (travelling at up to 10km/s or 6.2 miles per second); outward forces resulting from the habitats being pressurised for human survival;

radiation damage; lunar dust contamination; and the gravity that is one-sixth of that of the Earth. These lunar habitats will be a lifeline for future colonists and as such will have to provide oxygen for them to breathe, water to drink, an environment in which to grow food, protection from the harsh radiation of the Sun, as well as light, warmth and power during the 14-day nights. They will also need to keep people comfortable in all temperatures. There are two types of water on the moon: first, from water-bearing comets striking the surface; and second, originating on the Moon itself. These could provide a potential source of drinking water, fuel, breathable air and protection for inhabitants: it just needs digging up. In 2009, India's probe *Chandrayaan-1* discovered more than 40 permanently darkened craters near the Moon's north pole containing an estimated 600 million tonnes of water ice. Despite this huge potential store of water, over 90 per cent of that used in a future lunar habitat would be recycled. Recycled water would produce carbon dioxide (CO_2), which could be pumped into a greenhouse for use by plants that would in turn produce oxygen as a waste gas, which could then be pumped back into the habitat. Power requirements of a habitat would require a stable and continuous supply of energy that could easily be generated by lunar solar farms. On Earth, solar power generation is limited at night, but on the Moon there would be the option of 24/7 continuous clean energy generation (which could in theory be channelled back to the Earth, as well as being used on the Moon).

With such a low gravity compared to that of the Earth, building a habitat for living and working would become quite a feat. On the positive side, engineers would be able to build structures less hampered by gravity and moving large objects would be far easier; they would just need to make sure they kept hold of these objects. Conversely, the low-g environment would pose difficulties for construction workers and their ability to move around easily. The lack of an atmosphere, however, would prove most damaging.

Ignoring the obvious issue of a lack of air for inhabitants to breathe, without the buffering of air around drilling tools huge amounts of heat would be generated, causing drill bits and rock to fuse. Should demolition tasks be needed, explosions in a vacuum would create countless high-velocity missiles that would tear through anything in their path (including habitats and astronauts) and there would be no atmosphere to slow them down. Likewise, additional protection of habitats and inhabitants from meteorite impacts would need to be considered with no atmosphere to burn up incoming space debris. Also, ejected dust would obscure everything and settle statically, contaminating machinery, not to mention the huge health risk if the dust were somehow breathed in. The launch costs from Earth to bring building supplies would be astronomical, so local materials could and should be used wherever possible. Lunar regolith (fine grains of pulverised Moon rock), for example, could be used to cover parts of habitats to protect settlers from cancer-causing cosmic rays and to provide insulation. It is estimated that a regolith thickness of least 2.5m (8ft) would be required to shield the human body and reduce radiation exposure to a *safe* background level. High energy efficiency would also be required, so the designs would need to incorporate very effective insulating materials to ensure minimum loss of heat.

One design put forwards so far is a stereotypical inflatable dome, which would be lightweight and relatively easy to erect on the Moon's surface. However, this would need good protection against incoming space debris, micrometeorite attacks, solar radiation and the vacuum of space. Lunarcrete or mooncrete could be made from lunar regolith, water and cement with the cement manufactured from lunar rock with a high calcium content. Water would either be supplied from sources off the Moon or by combining oxygen with hydrogen produced from lunar soil. Another option could be to manufacture habitats via 3D printing. In September 2014, a 3D printer was sent to the *ISS* to help astronauts print tools, parts and other much needed supplies. For the Moon, ESA has a plan. Prior to a human mission, they propose to send in a shuttle with an inflatable dome that would be erected on the surface as the founding unit of a future base. A robot and 3D printer would also be sent to create an exoskeleton that would line the outside of the dome and provide protection. Using dust available from the surroundings, ESA estimates it would take three months to get the base constructed and secure for up to four astronauts to inhabit.

Habitats could also be erected within ancient lava tubes. These form when the upper layer of a basaltic lava flow cools and hardens, and molten rock continues to flow beneath it. Once this drains, it can leave behind a hollow tube-shaped cavity. These natural cave systems provide a structure within which habitats could be built and easily sealed, the rock itself providing protection from the harsh surface environment and impacts. Such lava tubes are commonly interconnected, which would provide scope for the habitat to grow. A tube 1km (just over half a mile) in size or bigger would be ideal. The presence of lunar tunnels has yet to be confirmed unequivocally but spacecraft have revealed cave entrances known as *skylights* that may open into hidden lava tubes. Because of the Moon's lower gravity, these are expected to be larger than those already discovered on our planet.

When prospecting for the ideal site for a lunar outpost, it should provide good conditions for transport operations, a range of useable natural resources and a number of targets of scientific interest. Yet, the success of a lunar settlement will heavily depend on the efficiency of its transport structure. It seems likely that transportation around the Moon will rely on wheeled methods, following from terrestrial vehicles and tried and tested *Moon buggies* from the *Apollo* missions in the 1960s and '70s. To avoid mission-ending dust issues, it would be necessary to construct roads or potentially even a lunar cable car. One lunar colony could utilise the Shackleton crater at the Moon's south pole, using the crater walls to enclose a domed city with a 1,520-m (5,000-ft) ceiling and a diameter of 40km (25 miles). A colony settled in that location would have access to large deposits of water ice and be situated on the boundary between lunar sunlight and darkness. Its proponents estimate that a Shackleton dome colony could support 10,000 settlers after just 15 years of assembly by autonomous robots.

Something to consider packing when moving to the Moon would be a good batch of microorganisms. Microbes are currently used in mining to help recover metals such as gold, copper and uranium as they can catalyse extraction of minerals faster than chemicals. In fact, more than one-quarter of the world's copper supply is currently harvested from ores using microorganisms, in a process called bio-mining. Microbes could also be an important food source on the Moon. They grow faster than plants and generate more breathable oxygen than the same volume of vegetation, and they have a simpler growing process and are cheaper to transport as they take up less space. A peanut butter and microbe sandwich, anyone?! On Earth, *Anabaena cylindrica* is a nitrogen-fixing cyanobacterium and an extremophile to boot. When tested on rocks on Earth that are similar to lunar regolith, this microbe was able to extract calcium, iron, potassium, magnesium, nickel, sodium, zinc and copper from the material. It could also survive for 28 days under the extremely low temperatures and pressures found on the Moon, as long as it was shielded from UV radiation. This useful organism could also tolerate a decrease in water availability, so could happily be freeze-dried for transport.

Humans may not have set foot on the Moon since *Apollo 17* over 40 years ago, but that doesn't mean other life forms won't grace the lunar surface again. NASA is teaming up with students and private space companies to grow the first plants on the Moon's surface. The self-contained *Lunar Plant Growth Habitat* will resemble a glorified coffee can and will contain enough water, nutrients and air to grow 10 turnip seeds, 10 basil seeds, and 100 arabidopsis seeds on the lunar surface – *Arabidopsis thaliana* (Thale cress) was the first plant to have its genome sequenced. This experiment will test whether plants can survive the radiation, flourish in partial gravity, and thrive in a small, controlled environment – the same obstacles that will need to be overcome in order to build a greenhouse on the Moon, and ultimately for humans to be self-sufficient on its surface. When the mini-habitat lands on the Moon, it will automatically release enough water to wet a piece of nutrient-laden filter paper. That, along with the natural sunlight on the Moon, should trigger the germination of the plants. Completely sealed, the container will only contain enough air for about one week, but that will be enough to show whether the seeds germinate successfully. Interestingly, as a control, which every experiment needs, NASA are crowd-sourcing by sending schools across the USA their own set of habitats so they can grow the same plants that are being sent to the Moon. If the seeds successfully germinate on the lunar surface, this will be the first terrestrial plant life transported to and grown on another planetary body. The experiment is destined to hitchhike on board the robotic spacecraft of whoever wins the Google Lunar X Prize, saving millions of dollars in travel costs. The current price tag rings up at a mere $2 million – quite modest for an experiment that could help us figure out how to sustain life on other planets.

In the meantime, during his year in orbit on board the *ISS*, Scott Kelly successfully grew red romaine lettuce and a flower in space – an orange zinnia. Kelly had agreed with NASA to tend the plant as if it were in his garden on Earth, rather than according to a strict scientific regime. The zinnia soon flourished and put out several buds: a thrilling indication that cultivation of crops for food and medicine may be possible beyond our biosphere.

A Mission to Mars

Should colonisation missions head for the Moon first or Mars direct? Although the Moon is closer, easier to access from the Earth and has near-instant communication options, Mars is the world that seems to have captured humanity's imagination as a future human outpost. Its formation and evolution are so similar to that of the Earth that we can't help but want to study it, to learn more about our own history and future. Robotic explorers have studied Mars for more than 40 years – it is the only planet inhabited solely by mechanised beings – but a human presence is still a long way off. Mars today, despite its sub-zero temperatures, thin, non-breathable carbon dioxide-rich atmosphere, high UV radiation and savage global dust storms, actually has, as we have already explained, the most clement and almost welcoming environment in the Solar System after the Earth.

At their closest point, a mere 54.7 million km (34 million miles) separates Mars from the Earth. One of the greatest barriers to humans making this trip, however, is the ability to carry the fuel needed. To travel these distances requires more fuel than just making a quick stop on the Moon, which means there is more weight to carry, and the greater the weight of a spacecraft the more fuel is needed to transport the weight. The total journey time from Earth to Mars could take between 150–300 days, depending on the distance between the planets at the time of launch and the rockets being used, so first on the to-do list would be to find a way to protect crews from radiation exposure on the trip. In May 2013, NASA scientists reported that a possible manned mission to Mars might involve a great radiation risk, based on the level of energetic particle radiation detected by the Radiation Assessment Detector on *Mars Science Laboratory* on its journey to Mars in 2011–2012. The *Curiosity* Mars' rover received around 0.66 sieverts during its 253-day cruise to Mars – the equivalent of receiving a whole body CT scan every five or six days. In addition to this, large solar flares and cosmic radiation, although they can be prepared for, may still deliver a lethal radiation dose to a crew. Nonetheless a human crew to the Red Planet would be likely to receive only just above the limit of radiation currently deemed acceptable over an astronaut's lifetime, and so the risk is judged tolerable in relation to the potential benefit to humanity resulting from their work.

Living in microgravity on the journey will cause the same physical effects as seen on the *ISS*, but these astronauts would then have to land on Mars and adjust to its gravity as quickly as possible. There will be no one waiting for them on the ground to help them out of the craft and to support them to start walking. The best way to assist the astronauts would be to remove the problem and produce artificial gravity by spinning the spacecraft as it travelled to Mars, adjusting the gravity slowly to help the astronauts adapt before arrival to life on Mars.

Another worry is the risk of supersonic space dust. In 1967, a stream of micrometeorites ended the three-year-long *Mariner 4* mission, while in 2012 a micrometeorite slammed into one of the *ISS*'s giant windows. Space shuttles have all returned to Earth with mini impact craters on their hulls. The *ISS* now has a micrometeorite shield on the exterior of the *Zvezda* Service Module, which any Mars shuttle would also need. Finally, although this is not the end of the list of worries by a long shot, there is the issue of landing. The current success rate for setting down safely and in one piece on Mars is only 30 per cent. With a human at the controls, the rate may be higher – NASA did not lose any of its *Apollo* landers during touchdown on the Moon. However, unlike the Moon, Mars has an atmosphere and its gravity can make a soft landing much more challenging.

Building an outpost on Mars will require a great deal more planning, even though we would have far better conditions to work with than are present on the Moon. The benefits of moving to Mars are that it has a similar length of day and the planet is tilted on its axis at an angle comparable to that of the Earth, which creates similar seasons; and it has an atmosphere, water ice, and habitable environments. There is a number of geological landforms, such as impact craters and lava tubes, which could be used to house habitats, as proposed for the Moon, and there is a never-ending list of scientific investigations that could be carried out. Larger habitats would be required for long-term living (the only real way to live on Mars), which could be built in stages during a series of launches, taking inspiration from the piece-by-piece construction of the *ISS*. The parts for a Martian base could be delivered by landing modules in a series of missions, which could then be assembled by a human crew either already on the surface or arriving later, by robots or even by a crew based on a nearby Martian moon. Buzz Aldrin, the second person to walk on the Moon and a great advocate of Mars colonisation, suggested sending three people to spend 18 months on the moon Phobos, using this as a headquarters from which to remotely construct a base for Mars.

The environment on Mars is the main challenge to overcome for any human or habitat as its 95 per cent carbon-dioxide-filled atmosphere is toxic to humans and promotes   low   atmospheric   pressures (0.006atm/0.0006MPa/6mbar). Additionally, it only has 38 per cent of the Earth's gravity, is always cold (−85°C to −5°C/−120°F to 23°F), and there are no liquid bodies of water on its surface. As for habitats on the Moon, oxygen would need to be generated for humans to breathe and suits would need to be worn whenever the inhabitants left the outposts. Owing to the time taken to travel between Mars and the Earth (not to mention the cost), a habitat on Mars would need to be self-sustaining from day one, growing its own food, extracting its own water and producing its own oxygen. Even though many studies are being conducted into the logistics of how this might be achieved, it is still very likely that a spacecraft would stay in orbit with food and supplies for a journey home, and also for a *safe haven* in case something went wrong on the surface. With costs of £50,000 to ship 4 litres (7 pints) of water to the Moon, imagine the cost, let alone the logistics, of shipping water and food to Mars on a regular basis. The only logical key to long-term human habitation of Mars is space agriculture or *astro-gardening*.

The first Martians will therefore be two species – plant and human – who are actually perfect travelling companions. Humans consume oxygen and release carbon dioxide. Plants return the favour by consuming this carbon dioxide during photosynthesis and releasing oxygen. Humans can use edible parts of plants for nourishment, while human waste and inedible plant matter can (having been broken down by microbes in tanks called bioreactors) provide nutrients for further plant growth. Plants need water, oxygen, sunlight, nutrients and relatively comfortable temperatures, but none of these are currently found on Mars in quantities that suit the growth of a garden. With a lower gravity than the Earth and ravaged by global dust storms, this is a world that would not support most known plant life. Plants are a canary in a coal mine for human habitation – if they cannot find a way to survive, then we cannot either. Gardening on Mars would provide a long-term food source for future human colonies and could provide over half their required calorie intake through the growth of tomatoes, potatoes and other fruit and vegetables. Plants that thrive in the carbon dioxide rich atmosphere of Mars include seeds of radish, alfalfa and mung bean, while asparagus, potatoes and marigolds have been shown to grow in Mars-like soils. If Mars gardeners are to use Martian soil, then knowledge of how crops respond to its contents, such as sulphates and perchlorates, will be required. Gardens help to recycle nutrients, and provide drinking water and in the longer term could provide building materials such as wood and bamboo. Any garden would need protection in the form of a greenhouse or geodesic dome that could keep the crops sheltered from extreme UV radiation, while still allowing in enough sunlight for growth. This dome would need to be securely anchored into the regolith to provide support and stability against the fearsome Martian dust storms and dust devils. The crops would also have to be kept warm while surrounded by the cold climate of Mars. This heating of the greenhouse would require energy, potentially from solar panels arranged outside the habitat and heating filaments beneath it; although that energy source too would require protection against the Martian environment. Accessing liquid water is a given since it would be needed both for irrigation of the plants and for human consumption.

Terraforming

When humans finally make it to Mars, we will live in a similar manner to how we do in Antarctica. Mini enclosed Earth-like environments will arise on the Martian surface that could include not only gardens and homes, but also parks, forests and lakes, all maintained under an Earth-like air pressure through a process known as paraterraforming. While it would cost a great deal of money to construct, paraterraforming sections of Mars with a sample of Earth's biosphere inside pressurised domes and underground caverns is something that humanity could achieve within mere years of arrival. Eventually, however, there would be an even more ambitious goal that might take millennia – full-scale terraforming. This is the process of 'transforming a hostile environment into one suitable for *human* life', as defined by NASA, although this is a dated and distinctly Homo sapiens-centric view. I prefer terraforming as the process of *making a hostile environment suitable for life* – not necessarily human life, which needs very specific conditions that are hard to achieve. Essentially, we are saying that one day we could bring Mars back to life. But why would we want to do this? Is it for the art of doing it, for the science, for the economic necessity or is it to leave a legacy for the planet Earth?

Terraforming Mars would entail three major interlaced changes: building up the atmosphere, keeping the planet warm enough to allow liquid water to remain stable on its surface and, finally, protecting the atmosphere from being lost to outer space. Most of the legwork would be done by life itself. You would not build Mars … you would just warm it up, throw in some seeds and allow life to take over. First, the atmosphere would need to be thickened and enriched with nitrogen and oxygen while the average temperature of the planet would need to be increased substantially. Perfluorocarbons, potent greenhouse gases, could be synthesised from elements in Martian dirt and air and then blown into the atmosphere. Through warming, the planet's frozen carbon dioxide in the ground would be released, which would amplify the temperature and boost atmospheric pressure to the point at which liquid water could flow. Terraformers might then seed the red rock with a succession of microbial ecosystems to increase the amount of methane in the Martian air, because methane is a much stronger greenhouse gas than carbon dioxide. First, bacteria and lichens (which have survived in Antarctica) would be grown, then later mosses. As dark plants and algae spread across the surface, they would darken the planet so that it absorbed more sunlight, and after a millennia or so, forests would be growing. With the right combination of plants and well-selected microorganisms, planetary engineers could generate the critical oxygen and nitrogen needed for human inhabitants to roam the surface without a space suit. Throughout this process, colonists would continue to inhabit Mars and expand the system of enclosed mini Earth-like habitats.

However, before we start this irreversible process of Martian environmental change, we need to be sure that we are not treading on the toes (if they have any) of any hidden life forms on the red planet. As we push Mars towards being more Earth-like, are there organisms present that might push back? We know that Mars has organic compounds that could be used by life, but a pessimist would say that any life that did exist there has not survived intact today. So if we are to bring Mars back to life, should it not be with native Martians rather than with Earthlings? The ingredients of the biosphere, if at all possible, should have Martian DNA at its core. Perhaps if we could find the relics of past Martian life frozen beneath the surface, in the polar ice caps, or living in some sub-surface refuge as we do on Earth, we could reconstruct it and let it once again control biogeochemical cycles on Mars. We could give Mars back its heartbeat. Sending life from Earth to colonise Mars should be an absolute last resort. Only if Mars has no genome should we consider sharing ours with it.

Epilogue

If the message has not been emphasised enough already, there are hundreds, most likely thousands, of possible habitable environments on worlds throughout our Solar System and beyond. But before we visit these places and possibly consider making changes to them so that they can host humanity, we may want to start by paying them the courtesy of protection from us while we investigate them. Space agencies all over the world have signed up to a very important programme of *Planetary Protection*, which scientists hope will preserve any possible life forms that do exist outside our planet from contamination by terrestrial organisms hopping a ride on incoming spacecraft. In turn, we are also committed to protecting the Earth from contamination by alien life should such a craft laden with samples return home. This responsibility is not taken lightly by humanity. We understand and appreciate the need to ensure that we are not interfering with these natural pristine environments while still powering forwards with our exploration of the cosmos. Missions to Mars are especially singled out by *The Committee on Space Research* (*COSPAR*). Since it has special regions within which terrestrial organisms could readily propagate, and areas thought to have an elevated potential for existence of Martian life forms. This applies to any region on Mars where liquid water occurs, or can occasionally occur, and is of course based on our current understanding of the requirements and conditions that are *just right* for life.

This talk of extraterrestrial life, habitable environments, moving to the Moon, humans colonising the Galaxy, and growing flowers and eating asparagus on Mars is fantastical and exciting, but we must not overlook and forget about our *just right* planet, to which we owe our very existence. We are designing habitats that will enable us to survive on

Mars and the Moon and technologies to search for alien life forms despite the fact that we do not yet understand how life came to be on the Earth and have been unable to colonise some of the most remote places on our own planet that are lush in comparison to those found throughout the Solar System. Humanity is not using the only truly habitable planet we know of to its fullest potential, but are instead taking a liberty with the lessons it has to teach us. Yet planetary exploration and planetary preservation have the same goal. If we can find means by which to colonise and inhabit the most inhospitable places on the Earth, it will aid our exploration of other worlds, and in so doing will help us to better use and preserve our own.

Whether we are searching for just the right home to live in, just the right school to send our children to, just the right moment to take the next step forwards in our lives, or just the right planet to house another form of life in the cosmos, humanity is constantly searching for its *just right*, as Goldilocks did. We are exploring the Solar System but found that Mercury and Venus may be too hot and Mars and the outer icy worlds may be too cold. We have eagerly turned our eyes to the exoplanets and exomoons of distant solar systems, hopefully watching and listening for any signs of life. Yet we have already found the *just right* for humanity and life as we know it today – the Earth – who knows what it might be tomorrow.

Acknowledgements

They say writing is a lonely and solitary process and yet I found it to be far from either of these. Although the act of putting finger to keyboard was of course done alone, behind the scenes I had an army of extraordinary people cheering me on, offering thoughts and suggestions, and sharing my excitement with the project. To start with I have to thank those without whom this book would never have been published: the incredible team at Bloomsbury Sigma and their fearless leader Jim Martin. Jim, you took a huge leap of faith by asking me to write my first book for you and I thank you from the bottom of my heart – your tweet could not have come at a better time. Thank you for always sharing in my vision for the book, especially its cover, and thank you for always replying to emails quickly and with a great sense of humour. To Sigma's assistant editor Anna MacDiarmid for taking *Goldilocks* through to publication, Lucy Clayton and to the rest of the Sigma team; this labour of love would not have been possible without you. The Bloomsbury Sigma family has brought me into contact with so many wonderful scientists and science communicators whom I am now honoured to call friends, and I am tremendously proud to be a part of this group. To my illustrator Sam Goodlet – I wish I had your talent! You are such a gifted and easy woman to work with. Thank you for sharing in my vision for the illustrations and I think everyone will agree you have produced some phenomenal drawings that bring the book to life. To Monica Byles, my brilliant copy editor – your comments have been invaluable and insightful and *Goldilocks* is the better for you.

A deep and heartfelt thank you goes to my huge support network of family, friends and trusted readers who have stood by my side and listened to my many ramblings about aliens and extremophiles: Natalie Bell, Fiona Bond,

Chloe Chin, Denise Houcke, Matthew Houcke, Tony Houcke, Jules Howard, Barbara James, Irina Kasatkina, Suzanne Kenny, Katherine Kotian, David Lambert, Kersten Malhan, Gemma Metcalfe-Beckers, Sophie Murray, Alexandra Pontefract, Ann Preston, Anke Schnedler, Linda Seward, Alaura Singleton, Lynne Whoolley and my TED Fellows family. I have enjoyed every second of writing this book and all of your support has made it possible. To those of you who gave up your time to read and edit my book – offering helpful suggestions, brutal comments and infantile humour over the use of certain words – this book is the better for you! I also owe a massive thank you to my favourite *coffice* (coffee office), Artisan, for patiently allowing me to steal a corner of its lovely cafe for hours on end. Thank you for the fantastic coffee, unshakeable Wi-Fi and delicious raspberry lamingtons, and especially to Jessica, whose steadfast optimism helped inspire me every morning. Thank you to Max Richter for providing the soundtrack to my mind and Netflix for letting me stream episode after episode of multiple TV shows, providing me with focus. Yes, I watch TV and write at the same time – I have no idea how or why – it just works.

To my parents, I say a huge thank you for … well … everything. Thank you for supporting and encouraging a seven-year-old girl who was just as interested in becoming a geologist and climbing volcanoes as she was in playing with Barbie dolls and running a delicious, yet make-believe, hamburger diner. Together and separately you have given me every possible opportunity to experience the best the Earth has to offer, and quite possibly sealed my fate in becoming a space scientist by orchestrating my first trip to the Kennedy Space Centre. Dad even masqueraded as a teacher to get me the Kennedy Space Centre educational pack – my hero. Throughout my life they have taught me to work hard and push my limits so I know this book would never have been written if it weren't for them.

Finally to Dan, my husband and best friend for the last 15 years, a thank you is simply not enough. Never was there a man who more supported a woman wholly determined to make a living doing a job she loved. Who patiently read her book on his commute to work every day and shared in her vision of the story she wanted to tell. Thank you for being my rock (pun intended) through the last few years and for giving me our wonderful little boy. I hope all the months of emotional and financial backing were worth it and I have made you proud. Oh, and I should also admit – he came up with the title – without him, there would be no *Goldilocks and the Water Bears*.

Index